Imperial College Inaugural Lectures in

Materials Science and Materials Engineering

U0319410

Imperial College Inaugural Lectures in

Materials Science and Materials Engineering

Editor

Don W. Pashley

Imperial College, UK

Imperial College Press

ICP

Published by

Imperial College Press
57 Shelton Street
Covent Garden
London WC2H 9HE

Distributed by

World Scientific Publishing Co. Pte. Ltd.
P O Box 128, Farrer Road, Singapore 912805
USA office: Suite 1B, 1060 Main Street, River Edge, NJ 07661
UK office: 57 Shelton Street, Covent Garden, London WC2H 9HE

British Library Cataloguing-in-Publication Data
A catalogue record for this book is available from the British Library.

ISBN   1-86094-106-0

本书由帝国学院出版社授权重印出版, 限于中国大陆地区发行。

PREFACE

Newly appointed or newly promoted professors at Imperial College are required to give an inaugural lecture on a subject of their choosing. The professors inevitably talk about areas of their subject on which they are expert, but the lectures are presented in such a way that they appeal to a wide ranging audience, from those who have little knowledge of the subject to those who are comparatively expert in the field. These written versions of six such lectures are aimed at readers who have some knowledge of, and interest in, materials science and engineering. They should also be of interest to those who are well versed in the subject.

The subject of engineering materials is highly interdisciplinary, so that research on materials is carried out in both science and engineering university departments and, in the case of Imperial College, also very much in the Department of Materials.

The six contributions in this volume are versions of some of the lectures given on materials subjects during the period 1993–1997. They include descriptions of the considerable progress made in some subjects, including major contributions by the inaugural lecturer, as well as interesting surveys of some subjects presented, in part, on a historical basis.

Many different materials are included, from metals and glasses to plastics and semiconductors. The practical use of the materials is covered together with the scientific understanding of their behavior. It is this combination which makes the subject of materials so fascinating and so rewarding for the researchers involved. There is scope for carrying out challenging science and, at the same time, contributing to practical applications in terms of the development of new or improved materials. The practical applications covered in this volume include: increasing the strength of cements and concrete, improving the processing of materials, the adhesive bonding of components in aircraft and cars etc., the construction of lasers and other solid state electronic and optoelectronic devices, bioactive materials for implants in the human body.

Don Pashley
Department of Materials

CONTENTS

Professor B.J. Briscoe

B.J. Briscoe was trained as a chemist and obtained a PhD degree in surface chemistry from Hull University, United Kingdom. He was Assistant Director of Research in the Physics and Chemistry of Solids Group at the Cavendish Laboratory in the University of Cambridge, and was also the Earnst Oppenheimer Fellow in Surface Science. Currently he is a Professor of Interface Engineering and Head of the Particle Technology Group in the Department of Chemical Engineering at the Imperial College, UK. His main research interests have included tribology and adhesion, ceramic and powder processing, gas-polymer interactions, suspension and paste rheology, interface mechanics, and nano-engineering. Awards and memberships include the Sir George Beilby Medal (Royal Society of Chemistry); the Society of Chemical Industry, Institute of Metals, for Advancement of Science in Practice; the Institute of Mechanical Engineers Tribology Silver Medal; and the Society of Chemical Industry Sir Eric Rideal Founders Lecture. He is an author or coauthor of more than 450 papers. He is currently Editor in Chief of *Tribology International*.

SLIPPERY CUSTOMERS; STICKY PROBLEMS

B.J. BRISCOE

Particle Technology Group
Department of Chemical Engineering
Imperial College of Science, Technology and Medicine
Prince Consort Road
London, SW7 2BY, UK
E-mail: b.briscoe@ic.ac.uk

1. Interface Engineering — The Scope

The term "interface engineering" is not in common usage but has received some interest as a new area of reasearch in recent years. It conveys the idea of modifying the interactions, or engineering the interactions between surfaces, usually solid surfaces, in order to provide profit and advantage. As far as I am concerned the subject had its roots in classical tribology which has an ancient history; the subject of tribology is defined later. In the context of my own work, I have chosen to combine the classical principles of tribology, which I learned at Cambridge and in my early years at Imperial College, with other types of engineering, in particular particle technology and more recently solid process engineering. To this group, we can add fabric engineering, geotechnics, magnetic media engineering and a new range of biotechnological applications.

Much of the work I have been involved with personally and most of the precedents involve engineering the interactions between solid surfaces, often by the interposition of organic materials, which are sometimes called "lubricants". Much focus is put on the practicality of such problems and this is why I have chosen the title "slippery" (lubrication) customers (the engineering context). The ideas of "sticky" and "problems" suggest that a very important component is the modification of the adhesion, but more importantly in the context of the current presentation, the friction, wear or damage of the solid surfaces. From the present perspective, there are probably four areas of important division of current interest, which are:

(1) Historical engineering tribology problems which could include the lubrication of engine components or the use of lubricants in a variety of bearings and machine components. Much of what I have learned personally, and is currently implemented, was derived from a general study of these types of problems but more importantly, from a detailed scientific study of the physics, chemistry and so on that is involved in the sliding of these solid macroscopic bodies.

(2) In recent years, probably beginning around the mid-eighties, it became fashionable and apparently valuable to incorporate these historical tribological principles into the description of what I would choose to call the description of "particulate assemblies".[1,2] The general idea is that the movement of particles over one another, be it in a salt cellar, where the particles are relatively dry, or in a mixture which involves solvent, perhaps in a paste, where there are latex particles and a solvent such as water, that the rheological or deformation behaviour of these systems could be described by a consideration of the interactions at the various contact points throughout the gross assembly. The idea then was that one might be able to develop, from a scientific base which could be derived the tribological principles, some means of interpreting formulation laws and indeed predicting *a priori* the rheology of these systems. This area has seen great progress in recent times in the development and rationalisation of the basis of formulation recipes in the areas of ceramic and food processing.

(3) Whilst classical tribology had dealt with the rubbing of relatively rough surfaces (see later), there had been precedents dating back to before the

beginning of this century for considering the rubbing of relatively deformable solids, such as red-hot steel in roll mills, in order to optimise the processing and forming of these types of materials.[3,4] Here, the contact areas might be relatively large and the deformations relatively gross. These two areas of historical tribology, those of relatively hard and rough surfaces and soft deformable surfaces, had not really met closely in terms of their scientific interaction until the consideration in recent times was given to the problem of what one now describes as "wall slip" in the processing of soft solids such as food pastes and ceramic suspensions.

(4) We can add another area, in addition to the ones described above, just to bring together some areas which do not naturally fit into the above descriptions. Here, we might be thinking in terms of the mechanical behaviour of a fibre reinforced polymer composite system or the mechanical response of human hair or synthetic fabrics. In these cases one is attempting to interpret the behaviour of the assembly, be it in terms of the tactile feel or the mechanical durability, in terms of the variety of interactions which occur within the assembly.[5]

So, in general summary, we would say that the Cause of interface engineering has a wide remit with a basic desire to develop means to modify the interaction, generally between solid surfaces, in order to optimise their processing and use. Although it would appear that the applications and contexts are extremely diverse, they do share a common scientific base in terms of the modification of the interaction between the solid bodies, be it at a gross scale, for example in an automobile brake,[6] or at a microscopic scale at, say, the junctions between the various fibres within a human hair.[7] It has been conventional to described these scientific subjects with the following generic headings.

1.1. *Adhesion*

Adhesion is the process whereby two solid bodies are brought into contact and, by a variety of mechanisms, energy will be required in order to separate them again. In the present summary, I will confine my interests to something which is called "contact adhesion" where there is no deliberate attempt to increase the adhesive strength of the junctions.[8] That other area will be well covered by our colleagues at Imperial College, in particular Professor Tony Kinloch,[9] whose interest is to optimise and understand ways of improving the adhesive strength between bodies, for example in the context of adhesive joints or in a composite system comprising of fibres and polymeric matrices.

1.2. *Friction*

Most of us will appreciate the general idea of friction and that it is the process of energy dissipation when two bodies are slid over one another or a force is required to initiate sliding.[10,11] In principle, there is no real distinction between adhesion which is separation with a normal force and friction which is separation whilst in contact by a tangential force. In the present discussion I will say little about adhesion *per se* and concentrate primarily on the role of friction.

1.3. *Lubrication*

Lubrication is a general term which is used to describe a process whereby a thin layer, usually a weak layer of a liquid or a weak solid, is interposed between the solid bodies in order to reduce frictional work and the damage of the solid bodies. There are a variety of ways in which this process can be achieved and the lecture dealt with some of these in the variety of the concepts which have been outlined above.[12]

1.4. *Damage*

Except at very low contact stresses and deformation rates, most solids would suffer some irreversible mechanical or chemical damage when they are slid over one another.[13] Often, the trick is to minimise this damage and hence minimise the wear (see Sec. 1.5). There are other circumstances, such as in abrasion or polishing or machining, where one would wish to maximise the damage. The difficulty in this area is that the damage processes are complex and are greatly influenced by the many system variables that can operate.

Little will be said here on this topic but it will be alluded to, in general terms, at the end of the present paper.

1.5. *Wear*

Finally, wear is a process which most of us would be familiar with where, through the action of sliding forces and the subsequent damage, mechanical erosion or chemical degradation of the solid bodies occurs. It is a frustrating and an expensive process and many of us spend a significant amount of our income and resources repairing the consequences of such damage.[14] For my own part, the most significant damage will be the replacement of my clothes, my shoes, my automobile engines or my automobile tyres. We can think of other examples.

So, the remit within the interface engineering context has many levels of requirement. At simple level, it might be to optimise the processing of a solid or the rheology of a paint suspension.[15] At a more diverse level, it could be to optimise the behaviour of a fabric conditioner[16] or a hair conditioner[17] or at a later stage perhaps wondering about our ability to minimise the wear and improve the friction of, say, automobile tyres.[18] The remainder of the paper will be divided into four or five general parts in order to exemplify the progression of my own interest in what I believe has been the way that this subject has developed in the last 30 years. The following section will deal with what I consider to be the very interesting historical context. The next section will describe some of the rudiments of the principles available for interpreting interface engineering problems. Two major sections will follow these, the first one being some examples of the tribology heritage in a classical sense of interfaces to cover such things as transport, cosmetics, ballistics. This is complemented by another section which deals with some facets of the interfaces within assemblies and the consequences upon the rheological response or processing behaviour of these materials. The final major section will deal with the topic of frictional walls and the consequence of interfaces in some aspects of the processing of solids, in the main starches and ceramic pastes. Finally, a brief discussion will be given to outline the challenges that still exist for the future and the new opportunities that may exist in such areas as bioengineering and nanoengineering in the context of magnetic storage media and so on.

2. Historical Contexts

"Tribology", as it is now called, has a long and rich history.[19] My own education here is largely derived from interactions with colleagues such as Professor David Tabor, but in particular the book by Professor Duncan Dowson, of the University of Leeds, called the "The History of Tribology".[19] Many of the transparencies that I used in my Inaugural Lecture and a good deal of the historical background was drawn from Professor Dowson's book and from monographs produced earlier by Professor David Tabor.[20] Here, we can record that, although the subject is ancient, the first records of any of major value can be found in the Egyptian hieroglyphics and the idea of improving the lubrication by the engineers who were responsible for the installation of the large statues in Egypt. Figure 1 shows such an example drawn from Professor Dowson's book. There it will be noted that the "tribologist" is holding a "jar" which he uses to pour liquid in front of the sledge in order to reduce the effort of the slaves who are pulling the sledge. Historically we had always believed that this jar had contained olive oil, but, in his recent inaugural lecture at Imperial College, Professor Hugh Spikes suggested that it was probably milk. He cited this as a good original example of using water based oil emulsions for lubrication purposes; environmental pressures now cause a move from oils to dilute oil/water emulsions.[21] The example is of interest because it defines the parameter, which is now historical, of a normalised frictional work or force and the parameter, the friction coefficient.

The friction coefficient which is now generally given the symbol, μ, which was originally chosen by Euller[22] in his thesis on friction, is defined as the ratio of the frictional force devided by the normal load in Eq. 1.

$$\mu = \frac{F}{W}. \tag{1}$$

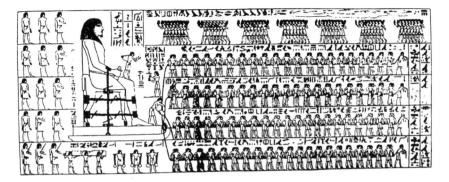

Fig. 1. The transportation of an Egyptian colossus from the tomb of Tehuti-Hetep, El-Bersheh, circa 1880 BC. This shows the lubrication engineer (see text) pouring "something" in front of the sledge to facilitate the sliding motion. Professor Dowson, in his book, makes the calculation that this corresponds to a friction coefficient of about 0.25. He estimated that there were 172 men each pulling with a force of 180 lb. The colossus is estimated to weigh 60 tons; see later.

The frictional force, F, can have a variety of forms. It can be the relatively constant frictional force required to maintain motion which is sometimes called the dynamic frictional force. Sometimes, because of instabilities, the frictional force will fluctuate and there will be such a thing as stick/slip motion.[23] We would be familiar with this in the context of the bowing violin strings or the motion of earthquakes. There is also an important friction which is called the static friction which is the force required to initiate motion and for many of us this will be quite crucial, for example when we begin to slip on a greasy bathroom floor. Invariably the static friction is greater than the dynamic friction; motion begins!

Many people believe that it was Leonardo Da Vinci who laid down the first "laws of friction"[24] which are quite familiar to most of us as they were taught to us in applied mathematics as school children. It was probably Amonton,[25] and subsequently Coulomb perhaps,[26] who formalised these "laws" in a form which we accept today. They are formulated in terms of the friction coefficient μ. The "Laws" suggest the following.

(1) The friction coefficient, μ, is independent of the load, W.

(2) The friction coefficient is independent of the apparent area of contact; see later.

(3) The friction coefficient is generally greater for rough surfaces than smooth and that "greasiness" will reduce the friction.

(4) The fricion coefficient (Coulomb[26]) is the order of 0.3.

Beginning in the 19th Century, many attempts have been made to provide a basis for the interpretation of these Laws and in particular the fundamental origins of these frictional process, the consequent damage and in more recent times the way in which lubrication can influence these processes. The idea that Coulomb had, which still has value, was that the friction arose from the engagement of what are called asperities. Coulomb's work, he was a military engineer interested in the launching of ships, was to do with the launching of ships down slipways; see Fig. 2. He had the idea, and it can be seen in his original works, was that somehow the frictional work was dissipated by the asperities moving up and down over each other or becoming engaged in such a way that might exist if one were to pull a hair-comb over one's fingers. There were concerns at that time, which were resolved to some extent by Bowden and Tabor,[10] about whether this type of process was actually energetically dissipative. Basically, "what comes up must come down" and the process as described had no fundamental dissipation mechanism. In more recent times, and this topic will be dealt with later, the Coulomb friction or Coulomb ratchet model has found value particularly where the surface roughnesses are comparable with the size of the solid body which can be the case with human hair or with powder systems. In more recent times, the belief in the origins of friction, and the next section deals with this in much more detail, was based upon of the fact that when solid bodies were put into contact and providing they were reasonably clean, then a significant adhesion, say welding, was developed at the asperities. Basically, the normal load was supported by the solid repulsive forces but when the solid surfaces came close together, at the areas of contact, as opposed to the total apparent contact area, strong adhesive junctions were formed. Upon the action of the sliding force these junctions were repeatedly

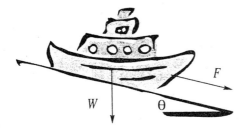

Fig. 2. A stylised representation of the ship launching exercise after Coulomb (see text). School text books tell us that the coefficient of friction, μ, which is the force, F, divided by the load, W, is simply equal to tan θ. An angle of about 20° is sufficient to launch most ships.

deformed and rupture during the process of shear. This immediately brings forward the idea that the contact between solids is rarely such that the real area is the same as the apparent area and that the extent of the roughness is an important contribution to the friction process. The late Philip Bowden[27] noted that, if one could look into the contact between relatively undeformable solids, as opposed to deformable solids which we will discuss later, then it would be such that one was taking the Alps and placing them on the Alps and the process of shear would be essentially at that sort of scale. In fact, we know now, using modern surface profilometry techniques, that a better example would be considering the South Downs sliding over the South Downs. There were a number of problems associated with the implementation of this model, which is now very widely used, but the main one was that how could one have adhesion in contact whilst sliding but not have adhesion when one attempts to separate the solids with a normal force. We do not stick to the floor; we can lift our feet off the floor by the action of a vertical force but we have to do a considerable amount of work if we want to slide our feet over the floor. The subtle intellectual trick, which was put together by Bowden and Tabor, was to consider that the solids were actually adhering but in the process of the adhesion in friction a significant amount of stored elastic strain around the asperities naturally parted the contact. This stored stress would not be dissipated during sliding but was dissipated during the unloading process. Quite simply the solids would peal apart. We can examplify this by taking a deformable ball, such as a squash ball, and "squashing" it against the table. There is no doubt that the surfaces are probably in adhesion because we would have to produce a significant amount of frictional force

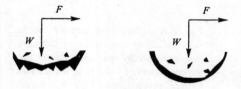

Fig. 3. A stylised representation of the two-term model of friction. On the left shows the adhesion component where surface asperities are sheared at the interface. The right shows the deformation term. The symbols are defined in the previous figure.

in order to slide. In fact, with such force that we might transfer, or wear by transferring, dark marks to the table. However, when one unload the body the stored elastic strains which are produced by the load naturally peal the interface apart and no significant normal adhesion is obtained. The next section will consider further what is now called the "adhesion model of friction". In the same period it had become of interest, primarily in the context of automobile tyres, to realise that there were frictional processes of a slightly different sort; see Fig. 3, where the friction could be dissipated, not by the shearing of interfaces, but by significant subsurface deformation. This became a major issue in the context of automobile tyres and the traction of tyres on wet roads. Many authors contributed to this topic and it could be argued that the topic is now largely solved but the following sections will introduce some of the simple examples of this problem and indeed the next section will describe how this process may be interpreted using fundamental principles.[3]

Ideas of the basis lubrication processes evolved progressively during the last Century beginning with definitive work by renowned surface chemists such as Langmuir[28] and Hardy[29] and many others. These people necessarily dealt with a type of lubrication which involved the deposition of relatively thin films, sometimes called monolayers, upon solid surfaces. The classic examples are usually attributed to Sir William Hardy who introduced the phrase "boundary lubrication". It was discovered by Hardy, and subsequently by many others including myself, that even a monolayer of a suitable organic solid, historically this would have been stearic acid, would significantly reduce the friction and damage between solid surfaces. A simple example is shown in Fig. 4 from my own work. The figure shows data for stearate. This data will be discussed at greater length in the next section.

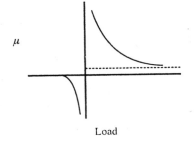

μ

Load

Fig. 4. The friction coefficient as a function of the normal load for a typical contact covering a wide range of loads. Under some circumstances, for fine contacts, it is possible to slide under a tensile load. The general trends of the friction coefficient at large positive loads are described later in the text.

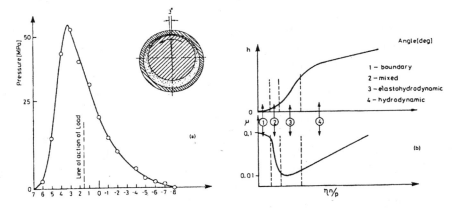

Fig. 5. Data taken from Tower (see text) and elsewhere. (a) shows the measured fluid pressure in a journal bearing as a function of position; also shown. (b) gives a schematic representation of the mean film thickness as a function of the Somerfeld parameter (see text). Also shown is the Stribeck–Hersey curve for the friction coefficient in the same coordinates. The various "regimes" of lubrication are defined.

It had been appreciated, in the latter part of the 19th Century, that effective lubrication could be achieved by interposing liquid layers between solid bodies in relative motion. The original definitive work was attributed to Beaucham Towers[10] who designed the lifting mechanism of Tower Bridge and the subsequent analysis which is now often attributed to Sir Osbourne Reynolds.[31] Figure 5 shows a simple example of the Beaucham Towers' experiment which involves a journal bearing rotating in a shaft. Also attached to this figure is a summary of what are now considered to be the important regimes of lubrication. The boundary lubrication mechanism was mentioned before: basically a solid lubrication process. The innovative Osbourne Reynolds process was called hydrodynamic lubrication, where the important parameters were the rheology of the fluid and the geometry of the contact, and to some extent, the normal load and the relative velocity of the sliding members. The analysis produced a result which suggested that the friction, which is shown along with the normal film thickness, would scale with the parameter viscosity times velocity divided by the normal load. In recent times, and the subject will not be discussed in great detail, considered another two forms of lubrication between the boundary and the thick hydrodynamic regime. There was the elastohydrodynamic regime which has been exhaustively treated in recent times, which involves a consideration of hydrodynamic effects and the elastic deformation of the solid bodies. The reader will realise that, although the loads may not be large, the contact areas could be small and hence very large pressures may be developed in the contact region. This has implications for the liquid or the fluid which would often produce a significant increase in its viscosity with pressure but also it will be sufficient to induce elastic deformation of the solid bodies. The paper will deal with an example later where such elastohydrodynamic lubrication processes are actually produced at relatively low loads in the case of the sliding upon the human body. For completeness, I should mention the regime of boundary lubrication plus elastohydrodynamic lubrication which is sometimes called the region of "mixed lubrication".

All these precedents are drawn primarily from what is now described as "Classical Tribology", and I have only dealt with the phenomenology here, which allows an interpretation, at least or a rationalisation, of the behaviour of a wide variety of solid contacts, be they the bearings of bridges, the lubrication of hair and people, ballistic fabric protection, or process engineering. The next section will amplify further, in general terms only, the principles which may be applied to the specific problems that I shall choose to address later. The details are not important; what is important is to demonstrate that there is a fundamental base, and probably a scientifically based predictive capacity, for interpreting the behaviour of a wide variety of systems and, what is more, the engineering of the behaviour to suit our needs. It will also be noticed that I have not dealt in this historical review with the arguably more serious problems of contact damage and wear. There are no laws, viable or accepted, of damage or wear and this area could be regarded as having rather little historical basis; it probably is one of the challenges for the future in the subject of interface engineering. The subjects of damage and wear will be dealt with briefly in the discussions of the challenges for the future of interface engineering.

3. Principles of Interpretation

The previous section, which described the historical context, has introduced some of the basic ideas which, when refined and quantified, provide viable predictive tools for the engineering of interface interactions. The cause in engineering is always to provide accurate predictions to achieve the necessary result and to minimise human labour and cost. Hence, wherever possible, the subject seeks to provide quantitative predictive tools. It will be seen however that the subject has not matured to such a level where this is always possible and, more often than not, only first order estimates are provided or, in retrospect, a sensible rationalisation of what has occurred can be made. In the latter case, this provides a better understanding and allows some predictions as to the operation of new systems. At a basic level, one would wish to predict the forces or the energies which are dissipated when solid bodies of defined character, sometimes with an interposed lubricant film, behave. For the purpose of this article, and indeed it was the case of the Inaugural Lecture itself, it is most convenient and reasonable to think in terms only of frictional forces and to neglect important key things such as contact damage or wear or perhaps, in the context of powders, comminution, or in fabrics surface damage and the loss of appearance.

In this context, it is useful to think of four sorts of frictional processes. These are exemplified in Fig. 6. In the figure four model contacts are shown and in each case the interpretation relies upon an approximation of the way in which the energy is dissipated during the frictional process. In

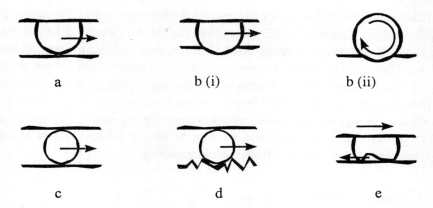

Fig. 6. The various modes of friction. (a) is the interface adhesive mode. (b) (i) and (ii) show the deformation mode both in lubricated sliding and in rolling. (c) shows the classical bulk deformation mode which may exist in certain types of assemblies. (d) shows the ratchet model after Coulomb which is typical of the response of hair systems (see later). (e) shows the phenomena of the Schallamach wave where a macroscopic dislocation propagates through the contact and no true sliding takes place.

the case of static friction there would usually be microslip or small movement between the contacting bodies before gross slip occurs and it would be useful to think in terms of work required to institute motion. There is also an additional problem in frictional processes, certainly in terms of fundamental measurements, regarding the mechanics of the machine. We know this to be the case in general studies; the motion of a violin string is an example. Here the response of the system is governed in part by the frictional response but also by the mechanics of the bow and the string. Often, this is a serious problem in translating fundamental data acquired from model experiments into the broader context of the large scale operation. The four examples shown can be grouped essentially into two or three groups. The first, Fig. 6(a) shows a representation of the "adhesion model of friction". Here, recalling above, the friction is argued to be dissipated by the repeated formation and rupture of adhesive junctions in shear induced by the natural adhesion at the points of contact. This model will be expanded upon below. The next example is the case, Fig. 6(b), which is commonly encountered in automobile tyres and some forms of human traction and grip.[18,32] In this instance, the frictional work is dissipated not in the interface shear zones or velocity accommodation loci, the term used by the French,[33] but in a gross subsurface deformation region. Here it is largely the plastic or dissipative viscoelastic properties of the material which is the important contribution to the frictional process. These are the first two major frictional processes as distinguished and accepted in recent times. An important first order approximation, which was invoked by Bowden and Tabor[10,34] and has been widely practised since, was that these two terms could be effectively added together and there would be no cross interaction of the two components. We will see later, in the context of certain processing operations, that this is not a good approximation and arguably it is fortuitous and of no value. It has however value where the contact areas are relatively small fractions of the apparent contact area and in particular where interfacial lubricants have been interposed at the contact zone. The third example is the case of rolling, a very common and widely used means of inducing relative motion between solid bodies. All of us would appreciate that it is easier to roll than slide and indeed this was familiar to ancients and the invention of the wheel was a significant improvement in the human engineering experience. The challenge of introducing rolling into bearings was developed over the last century with the introduction of ball and roller bearings and, of course, in recent times one has seen many facets of the way in which rolling can be produced with relatively little energy loss as compared to sliding.[35] The motor car tyre is an very good example of such and an interesting problem which will be cited in the next section.[18] The next example of a frictional process [Fig. 6(c)] bears some relation to the last but, in the context of the current interest, it has a certain novelty because rather than just thinking about a subsurface deformation being responsible for frictional dissipation, it is now largely the whole entity which is a dissipating body. Whilst this is not an important mechanism in ordinary tribology, although it is in process engineering, where it is commonly encountered in the rheology of relatively soft systems. For example, it is believed that the rheology of things like tomato ketchup is, in part, caused by the work dissipated in the deformation of particles as they move past each other. It is useful in the current context of interface engineering to consider this to be a frictional process also. In the fourth case [Fig. 6(d)], in this listing is the description of the so-called ratchet model proposed by Coulomb which was described previously as being generally inappropriate as a dissipative friction mechanism. This is probably the case in most large contacts but when the solid bodies become relatively small compared with their surface roughness then interlocking of asperities can produce certain interesting features and the example of the friction of human hair is one such example. Here, the reader may wish to perform a personal experiment if he has the necessary equipment. One simply grasps a single hair from one's head and stretches it and then one takes one's finger and rubs it up and down the hair. One will perceive that the friction coming down would appear to be greater than the friction going up. The reason is this, that the cuticles on the hair are grown very much like tiles on a roof. When one comes down the hair one is going against the tiles and there is more deformation associated with this process because of the ratchet mechanism as compared to when one slides down the "tile".[36] This is the basis of the shrinkage of natural fabrics. Figure 7 shows typical fraction data for human hairs. In the processing of solids such as particulate suspensions and pastes, this process can also be significant when the particle size is of the order of the wall surface topography.

In my own career in research, there have been many interesting periods, no more so than the discovery of the "Schallamach Waves". An interesting

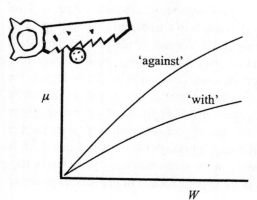

Fig. 7. A schematic representation of the so-called differential friction effect as observed for human hair. The friction coefficient is shown as a function of the load where two orthogonal fibres are slid over one another. Sliding against the "grain" or ratchet leads to a higher friction than when one slides in the "with" direction; see text. This provides the origin of the shrinkage of natural fabrics.

and probably unique phenomenon which provided a basis for a "new" friction mechanism for soft elastomers. The idea is shown in Fig. 6(e). I was privileged to spend a short period working on this problem with Andrew Briggs and David Tabor. Basically, we showed that friction could be explained in terms of viscoelastic adhesion by considering that macroscopic waves of detachment propagated through the interface; no true sliding takes place.[37,38]

In all of these cases it will be realised that the frictional force arises from and is dissipated by processes which involve the transmission of forces across interfaces. In order to make a start in defining and interpreting the behaviour of these systems, a number of simple quantitative models are required. In the present article only three of these processes will be dealt with in significant detail but there will be some overlap between the various deformation systems outlined in Fig. 6.

3.1. *The adhesion model of friction*

This model was introduced above and now finds wide favour and application in many engineering applications. It is now believed to be the mechanism of friction which was being investigated by Leonardo Da Vinci, Amonton and Coulomb and so on.[19,39] It is useful to exemplify its application and also that of the rationalisation of boundary lubrication, which is a corollary of this treatment, in the context of Coulomb's original experiments on the launching of ships. The Coulomb experiment and, indeed, a replica of the Leonardo da Vinci experiment but on a much larger scale, was shown previously. The idea is that at some critical angle of inclination on the plane the body will begin to slide down the plane. The important equation is shown; see Fig. 2.

$$\mu = \tan \theta. \tag{2}$$

Typically Coulomb and da Vinci found that angles of the order of 20° were ones which were necessary to induce motion and hence they believed that the friction coefficient was generally around 0.3. It will be recalled that Coulomb believed in the "Coulomb ratchet model" [see Fig. 6(d) above] but we can interpret quite simply the "laws of friction" as expounded in an earlier section using the adhesion model. Basically, the frictional force, F, is written as below in the following equation where the quantity τ is called the interface shear stress and A is the area of contact.

$$F = \tau A. \tag{3}$$

It should be realised at this stage that the contact area will, for rigid bodies (or bodies where we describe the asperities as being nonautonomous, but see later), be a small fraction of the apparent area; a ratio of 1000 would be quite typical. At this stage we should remark that in some facets of process engineering or in micro contacts such as in powders and hairs this would not necessarily follow and we will see there that the da Vinci or Coulomb laws do not apply. Thus we have many thousands of asperities each supporting part of the normal load and when the applied frictional force is imposed the interface contacts begin to shear, each with an individual shear stress and hence, if one sums over the discrete areas of contact, one has the net frictional force. The central problem in the application of this model when it was first devised, and it still is a serious difficulty, is the identification of the numerical values of the quantities τ and A and, in particular, how they may vary with some of the important variables in the system. In my own career I have focused very much on the estimation and measurement of the quantity τ, the interface shear stress. More of this is said in the next paragraph. The original idea was that the interface rheology for a clean solid, if such an animal exists in ambient, would be in some way related to the bulk shear yield stress of the material. I and my colleagues have spent many years using a variety of substrates, usually ones which are rather smooth such as mica or glass or specially polished metals, in order to avoid the difficulty in identifying the quantity A and setting up a model contact where we can compute or measure the area of contact, measure the frictional force, and hence compute the quantity τ.[40-42] The interest then is to understand how τ might vary with some important variables, of which the most interesting and possibly important is what is called the mean contact pressure, P. For simplicity we define here the contact pressure as the load divided by the contact area. If we do this then a rather common relationship shown in the next equation is observed experimentally.

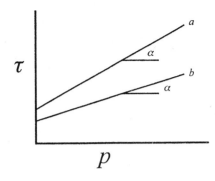

Fig. 8. The interface shear stress, τ, which is the frictional force per unit contact area as a function of the mean contact pressure which is the load divided by the contact area for typical systems. The text amplifies the case where the equation $\tau = \tau_0 + \alpha P$ may be used as a reasonably accurate description of the data.

$$\tau = \tau_0 + \alpha P \tag{4}$$

τ_0 and α are material properties which may be a function of several contact variables. Figure 8 shows such a relationship which is typical for a variety of organic films including pork fat, a few polymers and some so-called boundary lubricants. At this stage, although we will not advance the problem greatly in the current article, the quantity τ can be defined as a function of the sliding velocity or the contact strain rate or the ambient temperature or, indeed, it can be modified by the influence of various ambients such as water vapour and so on. Part of the interest here, and it will be expanded upon later, is to provide a means of understanding the variation of τ with the important contact variables. Dealing only with the pressure variable, we can take the above equation and the definition of τ and produce the following expression which indicates that the frictional coefficient is a function of the mean contact pressure for the system.[43,44]

$$\mu = \frac{\tau_0}{P} + \alpha. \tag{5}$$

The equation is interesting insofar as it demonstrates that the two important parameters for the interface, as far as pressure is concerned, are the intrinsic interface shear stress τ_0 and the pressure coefficient, α. The contact pressure arises from the nature of the deformation and the topography of the contact zone. Looking at the expression, we can see that if the contact pressures are large, which is often the case, then there is a prospect that the first term will be small compared with the second, the α term, and hence one will reach the situation where the frictional coefficient is just roughly equal to α.[39,45] In fact, this assumption need not be taken to this limit and there are ways,[46] where one can come to the conclusion that the coefficient of friction is just α plus another constant. In the context of the launching of ships using pork fat as a lubricant it turns out that the frictional coefficient was equal to the pressure coefficient of pork fat. Without digressing in detail about the significance of some of the principles involved here, we can quickly note that this relationship will then show that the frictional coefficient is independent of the apparent area of contact, independent of the load.

At this stage it is worth noting that what we have dealt with here is the use of adhesion model of friction, not in the context of the friction of clean solids but in the context of boundary lubrication. There are several reasons why one chooses to do this, not least the fact that this is the occurrence which usually exists in practice. Most solid surfaces are dirty and contain organic greases or are carelessly lubricated and hence one usually does accommodate the interface shear within a localised layer whose properties can be defined through τ in the way described. When the model was first put together by Bowden and Tabor and indeed others there were a number of refinements which allowed the model to be implemented for nominally clean contacts and in that case the interface rheology would be a characteristic of the solid. It turns out that this sort of model has problems in its application to metals but is quite widely applicable to organic polymers primarily because organic polymers generally create weakened interfaces through the action of the sliding process. So, for the case of an organic polymer the properties τ_0 and α are properties of the polymer surface and the property P is in some way often related to the bulk deformation properties and the topography of the contact.

Thus, we have a very simple model which requires experimental validation in order to interpret frictional behaviour, both classical and in the context of more advanced and interesting engineering problems. We note at this time that many refinements can be added to this sort of approach but essentially the value depends upon the significance and credibility of the principles that have been outlined in the previous paragraphs. In particular, there has been effort in recent times to attempt to predict *a priori* the interface rheological parameters from a knowledge of the chemistry or morphology or preparation procedures of the interface. In general, these tend not to have been particularly successful and one can quote here the attemps to optimise or maximise friction in certain types of contacts in the development of traction drives or polymer coatings.[47] Similar attempts have been made to optimise interface lubricant properties based on this model for fibre process engineering, hair conditioners and even skin conditioners. Our cause for the future would be to attempt to do more in this area. For our own part we have attempted to correlate infrared spectroscopic properties and interface rheological characteristics with limited success.[48]

3.2. *The deformation models*

We can now turn to the other major component of friction which is the so-called deformation friction model. Figure 9 exemplifies it in the context of rolling, but the same system response can be roughly achieved by sliding in the presence of affective lubrication. If one does not introduce lubrication there is significant traction or interface friction and the contact becomes severely distorted. Also in the presence of ineffective lubrication in rolling there can be significant contact adhesion between the bodies and the process of rolling can correspond to the repeated breaking and forming of adhesive junctions in rolling.[49] This problem was addressed by Andrew Briggs and myself some years and in this case good correlation can be obtained between rolling friction and the net pealing work. Figure 9 illustrates what the current

Subsurface
dissipation

Fig. 9. A schematic representation of the so-called deformation mode of friction where the frictional work is dissipated through subsurface dissipation of energy. The diagram uses as a force balance argument where the "push" is greater than the "shove" at the rear.

models attempt to describe and that energy is fed into the contact as it rolls, some is dissipated in the subsurface as elastic recovery occurs at the rear. One can think in terms of "pushing" against something but not getting "as much push" on the back or in terms of an energy dissipation argument the material undergoes a sinusoidal deformation during the rolling operation. A variety of models exist and the equation shown below is a reasonable description of these sorts of systems[3,10,20]

$$\mu = KE^{-1/3} \tan\delta . \qquad (6)$$

The point to note is that the major dissipation arises from the elastic compliance of the body although the term is small and represented by the Young's Modulus, E, but the crucial parameter is the viscoelastic loss characteristic, written here as $\tan\delta$, at the appropriate contact temperature and deformation rate. Figure 10 is an example from earlier work which demonstrates this principle.[50] Those of you who are familiar with Grand Prix racing and the excitement which accompanies the prospect of rain, will now probably realise the significance of the viscoelastic loss parameter in these sort of systems. When the road is wet the prime mechanism for grip is the viscoelastic defomation caused by the road asperities on the tyre. There is arguably no adhension and this is the only arrest mechanism. Hence a high loss rubber would naturally be required. This would give the necessary "wet grip". The same tyre in the dry would also have the contribution from the interface friction terms above and of course arguably the fact that there was high hysteresis loss would not matter. Unfortunately, the large hysteresis

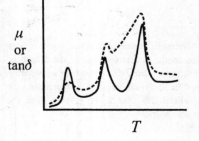

Fig. 10. The classical demonstration of the importance of the loss tangent in deformation friction; adapted from Ludema and Tabor. The friction coefficient, for the case of rolling on PTFE as a function of temperature, is shown (broken line). In addition, the loss tangent of the system at an appropriate deformation frequency as a function of temperature is also shown (solid line). The point to note is there is a correspondence between the mechanical loss and the deformation friction.

loss does infer that large amounts of energy is dissipated and hence the tyre would become hot. The tyre would then blister and fail. Our Japanese colleagues[51] have made many attempts to find ways of getting the best of both worlds in this context but it still remains a major challenge to engineer the appropriate mechanism based upon an environmental change such as rain. In recent times with a requirement to improve fuel economy there has been a tendency to use low-loss tyres, and indeed many trucks on their trailers use very low loss tyres. In this context you may wish to reflect upon the prospect of good braking resistance for these tyre types on a wet motorway!

3.3. *Other processes*

The next two mechanisms are not widely applicable in ordinary tribology but they are mentioned here for completeness. Referring above to the mechanism of subsurface defomation the idea here is that the deformation zone is comparable with the contact dimension. It is not supposed there will be very large deformations. However, in certain particulate systems to be exemplified later a very large part of the dissipation process within the assembly can arise not just from interface adhesive slip or just simple contact mechanical deformation but the gross deformation of the particles in the system. In these cases one seeks to develop models for describing the deformation of relatively soft particles. The argument is that the viscoelastic loss properties will be important and our own recent work has been to find an accurate measurement of the Young's modulus and the force displacement characteristics for the construction of rheological models for these systems. Figure 11 shows an example of the force displacement characteristics for a small particle. The particle is a swollen hydrogel which is often used as a model for investigating the rheology of gel systems.

The final aspect to be addressed is frictional engagement when the asperities are large compared with the contact dimensions. There are interesting areas here and much of our own fundamental work has actually capitalised upon this particular facet. In some cases for powders or indeed fibres, providing that they are reasonably smooth, when the bodies are brought

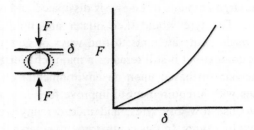

Fig. 11. Typical data for the compression of small spheres between rigid platens. The original sphere is deformed into ellipsoid in cross section. The deformation reaction is shown as a function of the imposed displacement for a typical case. Data adapted from Lui, Williams and Briscoe.

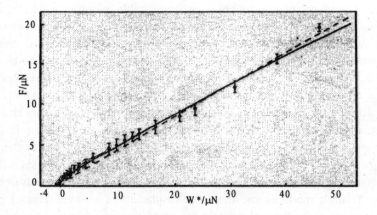

Fig. 12. An example of the friction of poly(ethyleneterephthalate) fibres against each other under low load shown as frictional force against normal applied load. The data points are experimental values for the stick friction and the full curve is a modification of the adhesion model of friction. Data taken from Briscoe and Kremnitzer.[52]

together the dimensions of the contact areas are relatively small and indeed comparable with the asperity size.[52,53] This allows simple contact mechanics for smooth surfaces to be used to compute contact areas and hence friction. The example shown in Fig. 12 is a case of the friction of two poly(ethyleneterephthalate) fibres in a specially devised machine. The data points are shown and the closed line is a prediction based upon the adhesion model of friction. Earlier, an example of the behaviour of stick-slip in fibre systems was shown where some of the kinematics of these systems is controlled in a large part by the topography of the surfaces.

In summary, we see that there are a number of notions or ideas of the origins of friction and some of the elementary bases for their quantification has been provided. Much more sophistication is available and continues to become available for providing predictions based upon available experimental data. It was recalled earlier that very little can be predicted *a priori* for these cases but this does depend upon the nature of the system under contemplation. The more serious issue, which is often not generally recognised, is that when one examines a system, and a variety will be examined later, is how does one know which of the four, or indeed there maybe more, models should be implemented for the description of the engineering of the system. The answer is that this is usually a matter of experience and this sadly limits the way in which the subject can be generally advanced and communicated. It has been said that tribologists, and I would now add interfacial engineers, are very good at rationalisation but have little genuine predictive competence. In terms of the current examples of frictional modelling the fundamental basis is actually quite reasonable compared with that which would exist for the cases of contact damage, or comminution or wear in more complicated environments.

The next section will deal with some more examples of tribological behaviour for macroscopic systems: classical tribology.

4. Tribological Interfaces; Some Examples

In this section I describe some of the work that I and my colleagues have been engaged upon in the area we have now begun to call "classical tribology". In this area, the major characteristic is that the solid bodies are large, although we will use some examples where they are relatively small. The important feature is that the asperities are relatively rigidly attached to the solid substrates which are in contact. In the vernacular of the subject we say that these asperities are nonautonomous.[54] In that sense it would be very much like sliding the South Downs upon the South Downs rather than sliding the Sahara upon the Sahara as a better analogy for the cases to be discussed in the next section. In the case of the former, the solids are relatively rigid and coherent and the asperities will be deformed in a cooperative manner. By this route, we would have discrete and separate areas of contact. When we slide the Sahara over the Sahara the expectation is that as the sand is relatively free-flowing we can relatively easily reach a situation where the real area of contact will be closer to, but still less than, the apparent area of contact. The latter case is dealt with in the final major section of this paper; it not amenable to a description using the original Coulomb laws of friction.

This section will contain three rather separate examples; the first is derived from the area of transport engineering or engine development. The second stems from what is now described as "cosmetic engineering", which will deal with hair conditioners and fabric conditioners and the way that these systems can be engineered to produce an acceptable tactile response. This part could arguably have been placed in the next section which deals exclusively with fine systems or with particle systems. The third section also deals with the fibre case and that is to do with ballistic armour protection. In each case the central idea is to modify or optimise and engineer the interface friction within the various elements of the system. The first example I use here, which fits in with the major lubrication theme described in the Introduction, is the lubrication of polymer/metal bearings within the context of hydraulic struts which are used in motor car applications. Many commercial front suspension units are constructed such that the suspension unit, which allows for the damping of the car under braking, also acts as one of the components within the steering mechanism. Such a system, which is now widely used within some manufacturing units, is called the McPherson Strut. In recent times it has been common to try and build the major bearing in this unit, that which allows the rotation for the steering and the vertical oscillation for the damping, around a circular metal strut and a polymeric bearing. A bearing which was fashionable in recent years is based upon poly(etheretherketone) and great hopes were placed upon this material to

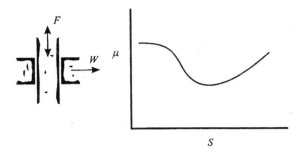

Fig. 13. Data for the lubrication of a McPherson's Strut, a component of a typical motor car suspension system. The frictional force, F, is monitored as a function of the side load on the bush. The shaft is made of steel. Typical data are shown for the friction coefficient as a function of the parameter, S, the Somerfeld number which is the product of the velocity times the fluid viscosity divided by the side load. These data are typical of those obtained in fluid lubrication in a wide variety of contacts.

replace the rather precious and environmentally difficult metals which had been used in the past. The ides is that efficient fluid lubrication should be produced and there should be very little environmental sensitivity within the closed environmental of the hydraulic fluid.[55] Figure 13 shows an example of the performance of such a system in a bearing simulator as obtained by Professor Stolarski (now at Brunel University) and myself some years ago. The important features are that the form of the curve follows very much the sort of behaviour that was outlined in the beginning of this paper which shows the transition from boundary through to hydrodynamic lubrication. Of particular significance here is the importance of the bearing tolerance in these systems and its influence upon the frictional behaviour.[56] It will be realised in these systems the frictional response is a crucial part of performance, not only to facilitate ease of steering, in spite of the prevalence of power steering mechanisms these days, but more importantly to ensure that there is no corruption of the natural damping behaviour of the strut assembly. The contribution of the interfacial engineering here is to design additive packages which can be included within the hydraulic oil, depending on the nature of the bearing system, which will ensure that relatively low frictions prevail throughout the course of the operation.[56] A key part of the difficulty is that when the relative velocity between the components approaches zero the friction is relatively high; there is no hydrodynamic film developed under these conditions. Professor Stolarski and I sought a number of recipes which might have provided a means of producing boundary lubrication for these systems. In the event, the work is now well reported, it turned out that the lubricants which had been widely used for metals could actually produce the converse behaviour for polymeric system.[55,56] In the event, we were unable to solve the problem from an engineering point of view but a scientific base provided confidence that the problem would not be readily sought using conventional chemistry and technology.

In most tribological actions, certainly in the large scale, the general aim has been to reduce friction. Reducing friction reduces energy consumption, damage and wear. There are, however, notable cases where this is not the case and that one seeks to have high and controlled friction.[57] The example of fluid lubricated traction drives was cited earlier and this still continues to be a major cause of interest amongst practising engineering tribologists. My own contribution in this area was to try and relate chemical constitution to the interface shear stress, although I must confess that what I had to offer had no real value in the final design of suitable lubricant packages. The most important example of maintaining high traction or friction is probably in the area of automobile braking systems or clutch drives. I worked for many years, I believe in a profitable way, with Dr. Paul Tweedale on trying to elucidate means of interpreting the friction of highly dissipative frictional contacts. At that time, our major preoccupation was the replacement of asbestos which was a key component in the construction of the common phenolic frictional components. Asbestos was regarded as an unacceptable environmental pollutant. With the generous help of one of our sponsors, Du Pont, we spent a lengthy period investigating the tribology of braking systems and trying to develop a methodology for describing the behaviour based upon simulations. Eventually, although we did not solve the details, we were able to provide an index of braking performance which is very much to do with providing a relatively constant frictional behaviour as a function of power dissipation. Figure 14 shows an example of the way we analysed this problem.[58]

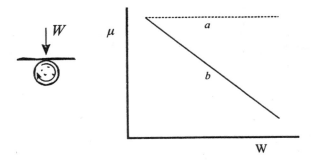

Fig. 14. Adapted schematic data for the friction coefficient as a function of the normal load for a brake simulator; data from Tweedale and Briscoe. Two types of data are shown. In both cases the frictional force and hence the friction coefficient is measured as a function of the increasing load in the simple experiment where the friction material rubs against a rotating shaft. Progressively the frictional heat increases the temperature of the system. The favourable case (a) is where the frictional coefficient is relatively independent of load but actually temperature. The unacceptable cases (b) corresponds to the phenomena of brake "fade" where the friction coefficient decreases with the usage of the brake material.

There has been quite a long history where people have sought to apply tribological principles to the behaviour of fabric systems and here I must acknowledge the many years of support that I have received from Unilever and Du Pont, America, in this cause. Our own work concentrated on a number of problems, which included the softening of fabrics and the generation of suitable hair conditioning formulations. In the latter period we have had good support from Proctor and Gamble and, more recently, from Unilever Research. Our major interests have been, and more will be said about this in the next section, to provide means of describing the behaviour, both adhesional and frictional, of individual junctions within a fibrous assembly so that we may have an opportunity, using a variety of means, to describe the deformation behaviour of the assembly of particles in the form of hair mass or the woven fabric. I must confess that we made relatively good progress in the context of the single fibre work and some of this was mentioned in an earlier section. In the context of the natural hair problem I can say that, with Professor Mike Adams, a longstanding research colleague, we were able to make I believe good contributions working on the frictional behaviour of rough natural fibre systems. Figure 7 describes what is called the diffferential frictional effects for a number of mammalian fibres.[59] The interesting thing, and it was described above, is the fact that friction is different in one direction as opposed to the other. There are some trivial matters here associated with why some fabrics shrink; the ratchet moves in one direction rather than another preferentially and how we might correct this behaviour to produce shrink-resistant fabric materials. The important contribution that we were able to make was to devise some models to interpret the origin of the frictional behaviour of these systems and to provide an opportunity to work on the behaviour of assemblies of such fabrics.[59] Although we made a number of attempts on various occasions, we have still yet to provide what I believe will be the definitive work on this subject. In more recent times, we have really rather concentrated on another form of what we would call cosmetic engineering, and that is the lubrication of the human body for a variety of uses. The example shown in Fig. 15 is based upon actual experiments conducted upon human bodies using a variety of lubricant media. In these cases the lubricant is relatively thick and hence we are not in the boundary lubrication regime but the hydrodynamic lubrication regime. It turns out that, as the human body is relatively deformable, it is probably the elastohydrodynamic regime (see earlier) which is responsible for the lubricating action in this area. Our analysis has actually shown that this is most likely to be the case.[60]

It is a pleasure for me, in this tribological component, to conclude with a brief description of work that was undertaken, in part with Dr. Daryl

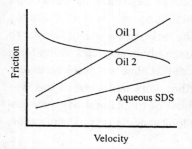

Fig. 15. The lubrication of homosapiens (inner arm). The friction is shown as a function of load for various lubricants. The sliding member is a glass sphere (see text). The oils are two silicone oils and SDS is sodium dodecylsuphonate (a common detergent). Oil 7 and SDS are effective elastohydrodynamic lubricants.

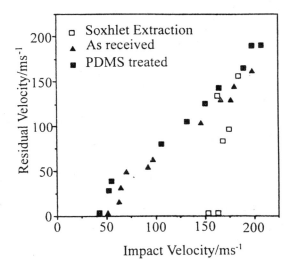

Fig. 16. Typical data for the ballistic capture efficiency of Kevlar© weaves. In the experiment a projectile is caused to impact upon a fabric which has had various forms of surface treatment (lubrication/cleaning). What is shown is the impact velocity as a function of the exit (residual) velocity. The point to note is that when the fabrics become contaminated or lubricated then less frictional work is dissipated during the capture and the projectile passes through the fabric at a lower impact velocity.

Williams and others, on describing the ballistic performance of fabric materials. This work was largely funded by Du Pont at Richmond, North America, and it allowed us to engage in the type of work which has been both valuable and interesting. The important thing in ballistic protection is to provide some sort of model as to how one might improve the behaviour of these systems. Figure 16 is an illustration of the high speed transverse ballistic impact data and shows the way that a fabric deforms during the impact of a projectile. In our own work we chose a quasi-static analogy and were able to provide a good account as to how frictional behaviour between the individual filaments and yarns within the fabric could be adjusted and modified in order to optimise ballistic projection efficiency.

These few examples, which could have been drawn from a wider net, provide a good prècis of some of the work that has been done by myself and my colleagues in the area of "classical tribology".

5. Interfaces in Assemblies

An assembly, in the current context, will be considered to be a large number of relatively rigid solid particles in a large volume. In some cases the particles will be suspended within a liquid medium and hence we have the prospect of moving from a dry powder mass through to a relatively dilute suspension. There are many variables to be considered in such a problem such as the particle size and aspect ratio as well as the particle volume fraction. In most cases the preoccupation will be to describe the rheology, in its most general sense, of the system and in particular the processing influence upon the final properties of the processed product. In this section it is convenient to discuss relatively dry powders in either compaction or in unconstrained flow and the behaviour of relatively dilute suspensions. The intermediate case, which has special problems, is discussed under the heading of "Frictional Walls" in the next section.

There are two questions to be addressed in such a study. One is the way in which the individual particles interact in a tribological sense and provide the overall rheological response. The interactions between the particles naturally produce the assembly flow behaviour but in the process of the deformation the structure of the system may also change. This is an important facet of the whole response of the system and its description and quantitative measurement still remain a major challenge. The second, and often neglected area, is the way in which the assembly interacts with the walls of the system. In macroscopic rigid bodies this would be considered to be a tribological problem but in the context of particle processing it is usually referred to as the "boundary condition". There are interesting differences between the topography of powder systems and that of monolithic rough specimens. The Downs versus the Sahara analogy was mentioned earlier. During the course of my research I have paid special attention to these problems as it is an area in which the now established principles of tribology may be readily

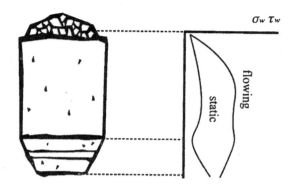

Fig. 17. Typical data for hopper flows from Tuzun, Adams and Briscoe.[62] Shown are data for the wall normal (σ_w) and shear forces (τ_w) for static and beds in flow. These data are well described by a modification of the adhesion model of friction.

incorporated into particle technology. It turns out that some of the problems addressed in tribology in the past have been relatively easily incorporated into the problem of interfaces in powder flows. The distinctions are relatively subtle and need not be dealt with at length here. However, an example suffices to introduce the opportunities here. The work I describe was undertaken with Professor Tuzun, now at the University of Surrey, and Professor Mike Adams at Unilever Research Port Sunlight.[62]

Figure 17(a) shows a pictorial view of powder flowing in a schematic hopper. A variety of powders were chosen for this study including glass spheres, poly(ethylene) particles and mustard seeds. A clever transducer system, developed by Professor Tuzun and his colleagues, was used to monitor the normal and shear stresses at the walls of the container whilst the powder was being filled and whilst it flowed. Figure 17 shows typical data for the normal and shear stresses either for the static fill or during flow. It is noticeable to see the important large normal stress, the so-called "switch stress", which develops at the point where the vertical side of the hopper transforms into the funnel which is conventional in these structures. The challenge in this work was to model, from single particle frictional principles, the sort of reaction forces which must exist at the walls during the flow of the powder. Typical experimental data and the associated theoretical predictions for the three particles are in good agreement. The important point here is that the theory, as developed by my associates and myself, was able to reasonably predict the behaviour of the actual silo system. It should be stressed that

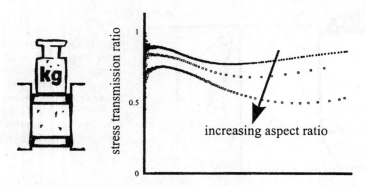

Fig. 18. A typical compaction cell where powder is compacted between rigid anvils in a die. Some of the applied force is dissipated as wall friction and this leads to the case where the applied stress is less than the transmitted stress; see next figure. Shown are the data for the stress transmission ratio as a function of the imposed stress. Various data are shown for different aspect ratios where peculiar trends are noted. The point of interest is that for the higher aspect ratios, taller specimens, the frictional losses are greater simply because the wall contact area is greater.

these data correspond only to the wall stresses and little is still known regarding how these stresses evolve as a result of the interactions between the individual particles in the flow in the hopper. Earlier work with Professor Adams and myself has demonstrated that similar principles could be used to describe the behaviour of powders at interfaces in rotating kilns.[63,64] In essence this work demonstrated, perhaps for the first time, that ordinary tribology principles could be usefully adopted, with some simple modification, to describe the behaviour of powders sliding against rigid walls.

Much of our recent work, however, has been preoccupied with "compaction flows" rather than unconstrained flows. Figure 18 shows a typical uniaxial compaction cell where the compaction stress is applied by the upper punch and the lower punch constrains the powder within the cylindrical die. Transducers attached to the upper and lower dies monitor the applied and transmitted stresses. Typical data, as a function of the compression on the upper powder surface is shown in Fig. 18. A relatively simple analysis, which is now over a hundred years old due to Janssen[65] and later modified by Walker,[66,67] allows the contribution of the dissipated wall friction work to be described in terms of the applied and transmitted stresses. The equation is relatively simple:

$$\frac{P_t}{P_s} \propto e^{\mu K \psi H/D} \tag{7}$$

where (see Fig. 1),[68] P_t is the transmitted pressure, P_s is the applied pressure, μ is the wall friction coefficient, K is the stress distribution correction factor, and ψ is a variable. H is the sample height and D the diameter, the ratio H/D is termed the aspect ratio.

By this route, a reasonably good estimate of the wall friction response was obtained. Recent work with Drs. Aydin[69] and Rough[70] have confirmed the general validity of the adhesion model of friction as applied to these sort of problems. I could note that with Dr. Andrew Smith and Dr. Marina Fernando[71] we were able to do this sort of work with maize powders over 15 years ago. More recently, Dr. Weert, Professor Adams, Dr. Lawrence and I have continued this work and developed it in the context of screw extrusion models for the compaction of powders.[72] The recent work in this area has proven, in conjunction with a major Link project which has involved a large number of contributions, that the simple uniaxial compaction process can be used as a good basis for providing the necessary materials response data for describing screw extrusion of powders and their compaction.[73] A more significant problem in the processing of ceramic systems is not simply the generation and transmission of stresses but the consequent complexities of the density distributions that are produced within the powder during its compaction. The natural censequence of the wall friction is to generate very complex stress fields within the powder and consequently significant variations in the density of the powder during the compaction process. For many systems, this is of no consequence but in the context of ceramics, which are sintered after compaction process, these variations in densities can lead to significant shape distortions of the system. An important aim in ceramic engineering, based upon dry powder processing, is to be able to produce specimens of a "near net shape form". In recent years Dr. Ozkan, Dr. Aydin, Dr. Sanliturk and I have developed a number of models based

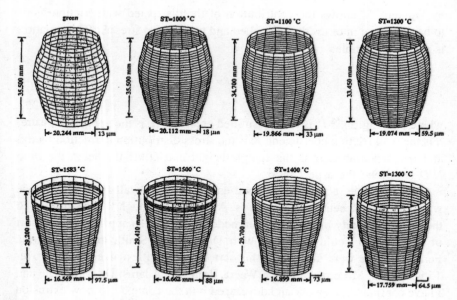

Fig. 19. An example of shape evolution during the sintering of ceramics which arises from nonhomogeneous density distributions within the compact (see text).

upon finite element techniques and refinements of those to predict the evolution of shape during sintering of cylindrical ceramic specimens.[74-76] At the present time we have generated a relatively good foundation to predict *a priori* the shape distortions of ceramic bodies during sintering. An example of the evolution of shape during sintering is shown in Fig. 19. Without going into detail, it is sufficient to say that the predictions based upon a variety of material input data, in particular wall friction data, have proven a good basis for predicting the observed sintered shape variations[76,77] using Finite Element methods.

Ceramic systems are conveniently processed by the dry powder route, as was outlined above. However, more commonly simple ceramic ware is made by producing a suspension, ideally an aqueous suspension, of the particles which are suitably stabilised by a variety of surfactants and polymer additives.

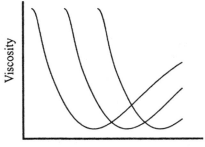

Fig. 20. Viscosity as a function of concentration for aqueous alumina suspensions containing various amounts of commercial dispersant agents. The point to note is that there is a critical concentration which produces the minimum viscosity and also the most effective processing route for these ceramic bodies. The viscosity scale is logarithmic whilst the concentration scale is linear.

An important part of this process is to ensure a good dispersion of the basic unit particles and to prevent agglomerate (caused by interparticle adhesion) formation during the manufacture of the suspension. These suspensions may then be dried or "slip-cast" and subsequently sintered in order to produce a useful ceramic product. Recently work with Dr. Ozkan, Dr. Khan and Professor Luckham has indicated clearly that a good suspension which is properly stabilised leads to the most highly dense fired product.[78,79] Figure 20 illustrates data. Recently we have completed a major EEC Scheme, which has involved about ten laboratories, which has sought to optimise and identify the important scientific base for the formulation of these sorts of ceramic suspension systems in order to optimise ceramic component manufacture.

6. Solid Processing and Production Engineering; Frictional Walls

Much of what has been described previously has been based upon two tenets of available topographical principles. The earlier work has described the interaction of rough surfaces where the asperities are securely attached to the supporting solid bodies. This leads to most interesting phenomena such as the classical laws of friction. For powders of the sort that was described in the previous section these laws are relaxed because the particles, which are now the asperities, are not firmly attached to the solid body and have what was called asperity autonomy. These lead to very interesting deviations form the "classical tribology" based laws of friction. In the context of the dilute particulate suspensions it is normally supposed that there is no wall slip between the suspension and the wall. An important intermediate case, which is now attracting very wide attention in the national and international research community, are the cases where a relatively small amount of liquid is incorporated into a relatively large volume fraction of powder. These are the so-called soft solid or paste systems. There is debate as to what actually constitutes a "paste" but for the present purposes it is a two-phase system where about half or more of the system comprises a solid particle; the rest is a fluid phase. The deformation and response of the system, which is controlled in part by the boundary as well as the interparticle motion, arguably corresponds to a special case of the lubrication problems described previously. The actions of these lubricants are very similar to those which were encountered by Osbourne Reynolds and Beaucham Towers.[80] The study in this area has not advanced sufficiently to make the intimate connections between the two cases but we have made some progress in recent years to at least interpret the subtleties of the behaviour of the boundary between the paste and the wall.[81] This is an important problem in the design and operation of such things as screw extruders, roll mills and extrusion units.[82] The important consideration which has evolved through our most recent study in the context of food processing, under the auspices of a major MAFF/DTI Link[83] Scheme, was to identify the ways in which the interface can be reasonably and sensibly described. There are important questions also regarding the intrinsic bulk rheological response of these systems. Much of our work has been concerned with food, particularly granular starch-water systems, but we have also been involved in ceria-gadolinia systems for the production of solid oxide fuel cells. Basically, one requires a set of constitutive relationships in order to describe the bulk response as well as the interface boundary (friction) condition behaviour. The equation shown below in terms of an interface wall shear stress as a function of the key operational variables seems to be an adequate description within the accuracy of the current experimental methods from the systems that we have addressed.

$$\tau = \tau_0 + \alpha P + h \frac{V}{V_0}. \tag{8}$$

We note the similarity with an earlier expression, Eq. (4), for describing adhesive friction and boundary lubrication. V is the velocity, and V_0 are constants.

It cannot be stressed too highly how important the role of the wall boundary condition is in the energy dissipation and mixing processes for these systems. In recent work Dr. Corfield, Dr. Sinha, Dr. Aydin, Dr. Lawrence, Professor Adams and myself have spent significant effort in developing means to identify and characterise the important response features in these systems. Our preferred routes have been based upon what are called the classical upsetting methods,[84,85] from hot metal working methodology and from the capillary extrusion based upon precedents laid down by John Bridgwater and John Benbow.[86,87] Figure 21 shows some typical data for upsetting which illustrates the important role of the wall boundary condition in the power dissipation during the deformation of a paste-based system. In the context of applying these data to screw extrusion, Fig. 22 shows data[88] which has been used to predict the behaviour of a food paste (potato starch) in a single-screw extrusion system based upon measurements made from capillary extrusion.

Before concluding this section it should be noted that the precedents here are developed from plasticity analysis; it is supposed that these

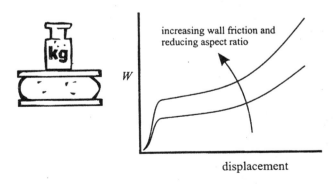

displacement

Fig. 21. Examples of upsetting where a deformable plastic cylinder (paste) is deformed between rigid platens. As the friction increases the deformation force increases.

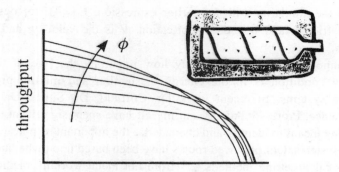

Fig. 22. An example of the prediction of output for a screw extruder moving compacting powder as a function of the conveying angle, φ for various pressure gradients, P.

paste-like materials, like ductile metals, have an intrinsic yield behaviour. However, three points are worth mentioning here. Firstly, since the solids are soft and extensively deformed there is little asperity persistence and contact areas reach apparent contact area values. This clearly simplifies the analysis although it does beg the question as to what is a contact. Secondly, it will be realised that now there is a strong interaction between the interface stress and the bulk deformation stresses; as a consequence the adhesion model of friction described previously is not applicable. If would seem that a route based upon a fluid dynamics approach, which was the basis for the Reynolds treatments which are introduced earlier, might be helpful. Finally, it is interesting to reflect now upon the differences between the classical tribology approach for rigid surfaces, the behaviour of powder walls and the behaviour of soft solid walls and this is the final reflection of this essay.

7. Final Reflection

In the Introduction a variety of frictional models were introduced. Particular emphasis was placed upon the so-called "adhesion model" and the "boundary lubrication model" and its implications as far as lubrication engineering is concerned. Some reference was made to rolling friction and also deformation and ratchet friction. A major mission in my professoral life has been to incorporate classical tribological principles into powder technology and soft solid process engineering. The key issue has been, and still is, as to how the two can be properly merged in order to provide a unique common basis for the advancement of the science of both. It turns out that the essential difference is that particles, be they dry powders or pastes or dilute suspensions, differ from macroscopic or monolithic rough solid surfaces only because of the difference in the autonomy (contact-mechanical relaxation) of the contacting members. For rough relatively undeformable solids the contact area is a relatively small fraction of the apparent contact area. As we recalled earlier, this provides a good interpretation of the classical Coulomb or Amonton laws of friction. Basically, the frictional force is dissipated at interface asperities and these asperities are rigidly connected to the bulk of the adjacent material. Interesting new contact mechanics models arise here which extend upon the work of Greenwood and Williamson, and Archard which provided an interesting and now well rehearsed basis for interpreting the friction of ordinary solid bodies. When one moves to powder and indeed paste systems then one quickly reaches a situation where the apparent area of contact is actually close to the apparent contact area. The powders can move at the interface either because of the movement between the dry powders or within the pastes and dilute systems to ensure that the contact area is very close to the apparent contact area. In principle, the treatment of these problems is not greatly different to previous cases and arguably it is a simpler case also. One of the challenges for the future will be to incorporate the precedents of lubrication theory, specifically hydrodynamic lubrication into the treatment of pastes or dilute systems. At present we search for means of describing these systems. Our new research initiatives are looking at means to identify the way in which fluids migrate to the walls of pastes during flow.[87] It is perfectly possible that we will be able to incorporate lubrication approximations in order to describe the rheology of the walls of these systems.

When I finally completed the basic structure of the present manuscript, I reflected that many important contributions to our understanding of various tribological, processing and interfacial phenomena had been developed outside of the references and structure which is described here. It is worthy to note this because it demonstrates the important breadth and interdisciplinary nature of interface engineering. I would mention particularly at this time the work I undertook with Dr. Panesar and Dr. Khan on the use of friable particulate layers as mould release agents.[90-94] In addition, in recent times my colleagues and I have spent a significant amount of time looking at the scratch or deformation behaviour of polymer surfaces and our interactions, previously with Dr. Evans and Dr. Sebastian[95-97] and more recently with Dr. Pelillo, have led to much better understanding of the durability of surfaces. I should also mention the extensive studies that I and colleagues have undertaken where gases are caused to imbibe into polymers. This work was originally undertaken to understand the behaviour of gas compressors which contained polymer piston rings. The work evolved into many areas now which include the foaming of plastics, the foaming of food and the study of the explosive decompression failure of elastomers. In this area I particularly acknowledge the contributions of Dr. Mahgerefteh, Dr. Liatsis, Dr. Savvas, Dr. Kelly and Dr. Zakaria who contributed in this area.[98-101] I and colleagues, initially some years ago with Professor Klein, studied extensively the migration of additives and plasticisers through polymers as a means of providing internal lubrication.[102,103] This work has led on to all sorts of innovations in terms of surface modification of polymers, surface plasticisation and other phenomena. Here, I should acknowledge the many contributions in the area of spectroscopy provided by Barbara Stuart and Paul Thomas.[104-106] I also recall extensive studies that we have undertaken looking at the condensation and wetting of polymer surfaces, primarily in the context of improving fluid lubrication. Many interesting mathematical theorems were proved in the course of this work and I do acknowledge the many contributions provided by Kevin Galvin, now of the University of South Wales.[107-109] The general theme of particulate arrays and their durability continues to be explored in the context of granulation by Dr. C. Lawrence, Professor P. Luckham, Professor M. Adams and myself. This work had its roots in the studies of mould release coatings and more recently in developing protocols to enhance the efficiency of the cleaning of plant[110] and the rheology of drilling muds and polymer solutions under high pressure.[111,112] I do, however, reflect, that the work which drew the tribology and procesing strands together, was with Dr. R. W. Nosker, then of RCA America, where we studied the role of interface friction during the deformation of polymers at high rates of strain using a split Hopkinson pressure bar. It was there, that I first realised the crucial influence of interface friction in material deformation.[113,114]

These examples, which is not a comprehensive listing, I would hope demonstrate the variety and richness of the subject of interface engineering. There are a number of ways to engineer the properties, particularly mechanical and chemical properties of interfaces, to produce desirable results. In the context of engineering, one was always looking for the most cost effective and durable solution. The fact that we have chosen to explore such areas as the diffusion of gases and low molecular weight species in polymers or indeed the trapping of particles on polymer surfaces simply demonstrates that the problem is complex but, because of its complexity, is amenable to fine tuning. Interfaces are complex things and deserve a detailed examination.

8. Challenges for the Future

One can reflect that the precedents laid down by da Vinci, Coulomb, Amonton and others which have been amplified by the work of Hardy, Dowson, Tabor, Bowden and so on provide a secure basis for the prediction of the friction of rigid solids. The present essay has not dealt with some of the consequences associated the consequent damage and wear of these systems. This is a much more problematical area but deserves our special attention for the future. The translation of these notions into process engineering of powders and pastes or indeed dilute suspensions is an interesting prospect and also a major challenge for the future. It is readily appreciated that the lubrication or slip at the walls of paste systems could be interpreted based upon the Reynolds and Dowson EHL arguments. We have yet to define means of doing this. If the walls are rough then it is possible that the deformation friction mechanisms as introduced by Tabor and Johnson and others could also be applied to the wall boundary in these systems.

I feel personally privileged to have been involved in two facets of engineering research. My days in Cambridge provided me with a unique introduction to the tribology of rough monolithic surfaces and introduced me to the important notions of topography, wall slip, friction, damage and lubrication and so on. In the context of engineering, as a chemical engineer at Imperial College, I found application for these ideas in a variety of processing operations. Our simplest successes have been in the areas of hoppers and in compaction. We look to the future in order to able to develop means of using these ideas in the context of examining the details of the interface slip condition for paste and dilute systems. Our major challenge ahead will be to introduce these ideas into describing the interaction between individual particles within, say, a paste flow or a powder flow. This is a much more significant challenge and one also which will require the introduction and development as well as implementation of a wide variety of numerical techniques. We have spent several years using finite element analysis methods and distinct element analyses are currently being implemented. The major difficulty is that the interaction laws which we seek to develop by a variety of routes changes the structure of the system and the structure changes the interaction and hence the overall rheological response both in the bulk and at the walls. We might consider this area of internal structure prediction and sensing to be our major challenge for the future; in simple terms a study of "damage-induced" changes. A proper understanding of these interactions and how they can be engineered will provide a means to either rationalise formulation laws or to predict *a priori* the necessary ingredients and interactions which are required in order to convey the appropriate processing conditions in order to make optimal materials based upon particulate entities. This would be very much in the best spirit of engineering.

9. Personal Remarks

This article was prepared almost seven years after the presentation of the Inaugural Lecture whose title it bears. The basic elements and themes of the lecture and article are similar. However, many of the examples are taken from more recent work.

During the preparation of both the lecture and the article I have been continually reminded of the many colleagues and students who have contributed to both. I have named some in the text, but be assured that I have enjoyed, and in some cases continue to enjoy, our interactions and achievements. I thank them for allowing me that special privilge.

Acknowledgements

I am pleased to acknowledge the support and input of very many people during the preparation of the original inaugural lecture and this present paper. I should also recognise the many contributions to the overall studies that have been provided by my students and research associates. Particularly worthy of mention at this time are those people who helped me, in a difficult period, to produce this manuscript. I am particularly grateful to Mrs. Joyce Burberry and Dr. Ismail Aydin for producing the manuscript and Dr. Xavier Weert for the generation of the figures.

References

1. B.J. Briscoe and M.J. Adams (Eds.), Tribology in Particle Technology, Adam Hilger, 1989.

2. B.J. Briscoe, *Powder Technology*, **88**, 255, 1996.

3. K.L. Johnson, Contact Mechanics, Cambridge University Press, London, 1987.

4. G.W. Rowe, Elements of Metalworking Theory, Cambridge University Press, 1986.

5. J.W.S. Hearle, J.J. Thwaites and J. Amirbayat, Mechanics of Flexible Fibre Assemblies, Sijthoff and Noordhoff, USA, 1980.

6. B. Ramirez, P.J. Tweedale and B.J. Briscoe, I. Mech. Eng., *Proc. Disc Brakes Commercial Vehicles Conf.*, London, 1988.

7. C.R. Robbins and C. Reich, *J. Soc. Cosmet. Chem.* **37**, 141, 1986.

8. D. Tabor, *J. Coll. Interface Sci.* **58**, 2, 1977.

9. A.J. Kinloch, Adhesion and Adhesives: Science and Technology, John Wiley, New York, 1987.

10. F.P. Bowden and D. Tabor, Friction and Lubrication of Solids, Parts I and II, Oxford University Press, 1950, 1964.

11. E. Rabinowicz, Friction and Wear of Materials, John Wiley, New York, 1966 and I.V. Kragelskii, Friction and Wear, Butterworth, Washington, 1965.

12. B.J. Briscoe and D. Tabor, Lubrication in Interfacial Phenomena in Apolar Media, eds. J. Eicke and G. Parfitt, Marcel Dekker, New York, 1985.

13. R.D. Arnell, P.B. Davies, J. Halling and T.L. Whomes, Tribology: Principles and Design Applications, Macmillan, 1991.

14. M.J. Neale (Ed.), The Tribology Handbook, 2nd Edition, Butterworth Heinemann, Oxford, 1995.

15. R.A. Williams (Ed.), Colloid and Surface Engineering: Applications in the Process Industries, Butterworth Heinemann, Oxford, 1992.

16. B.J. Briscoe, A.I. Bailey and S.A.R. Sebastian, *Textile Res. J.* **56**, 604, 1986.

17. B.J. Briscoe, M.J. Adams and T.K. Wee, *J. Phys. D. Appl. Phys.* **23**, 406, 1990.

18. T. French, Tyre Technology, Adam Hilger, Bristol, 1989.

19. D. Dowson, History of Tribology, Longman, London, 1979.

20. D. Tabor, Friction, Lubrication and Wear, *Surface Colloid Sci.*, ed. E. Matijevic, John Wiley, New York, **5**, 245, 1972.

21. H. Spikes, Inaugural Lecture, Imperial College, London, 1997.

22. L.L. Enher, *Mém. Acad. Berlin*, **4**, 133, 1748.

23. B.J. Briscoe, D.C.B. Evans and D. Tabor, *J. Coll. Int. Sci.* **61**, 9, 1979 and *Proc. Roy. Soc. London*, **A380**, 389, 1982.

24. E. MacCurdy, The Notebooks of Leonardo da Vinci, Jonathan Cape, London, 1956.

25. G. Amontons, *Memoires de l'Académie Royal*, (Chez Gerald Kuyper, Amsterdam, 1706), **A**, 1099, 257.

26. C.A. Coulomb, *Mém. Math. Phys.* (Paris), **X**, 161, 1785.

27. D. Tabor, private communication.

28. I. Langmuir, *J. Am. Chem. Soc.* **39**, 1917, 1848 (see also J.N. Israelachvili and D. Tabor, *Nature, London*, **241**, 148, 1973).

29. N.B. Hardy and I. Doubleday, *Proc. Roy. Soc. London*, **A104**, 25, 1923.

30. B. Tower, *Proc. Inst. Mech. Eng.* 50, January 1865.

31. O. Reynolds, *Phil. Trans. Roy. Soc.* **177**, 157, 1886.

32. M.J. Adams, B.J. Briscoe, S.A. Johnson and D.M. Gorman, Elsevier Tribology Series, 23, *Proc 19th Leeds-Lyon Meeting*, Elsevier, Oxford, 57, 1992.

33. D. Dawson, C.M. Taylor, M. Godet, D. Berthe (Eds.), Tribology Series 12, *Interface Dynamics*, Elsevier, Holland, 1988.

34. B.J. Briscoe, *Phil. Mag.* **A43**, 511, 1981 and Fundamentals of Friction: Macroscopic Origins, eds. E. Singer and H. Pollock, Kluwer Academic Press, Netherlands, 1992.

35. T.A. Stolarski, Tribology in Machine Design, Industrial Press, London, 1990.

36. B.J. Briscoe, M.J. Adams and A. Winkler, *J. Phys. D. Appl. Phys.* **18**, 2143, 1985.

37. G.A.D. Briggs and B.J. Briscoe, *Wear*, **35**, 357, 1975.

38. I. Shallamach, *Wear*, **17**, 301, 1971.

39. B.J. Briscoe, Friction, Fact and Fiction, *Chemistry and Industry*, 467, July 1982.

40. B.J. Briscoe and D. Tabor, *J. Adhesion*, **9**, 145, 1978.

41. B.J. Briscoe, B. Scruton and R.F. Willis, *Proc. Roy. Soc. London*, **A353**, 99, 1973.

42. B.J. Briscoe and A.C. Smith, *J. Appl. Poly. Sci.* **28**, 3827, 1983.

43. B.J. Briscoe and P.S. Thomas, *Wear*, **53**(1), 263, 1992.

44. B.J. Briscoe, Interface Friction of Solids, *Proc UK and India Royal Society/ Unilever Fora*, eds. M.J. Adams, S.K. Biswas and B.J. Briscoe, Imperial College Press, London, 1996.

45. J.K.A. Amuzu, B.J. Briscoe and M.M. Chaudhri, *J. Phys. D. Appl. Phys.* **9**, 133, 1976.

46. M.J. Adams, B.J. Briscoe, M. Streat and F. Motamedi, *J. Phys. Appl. Phys.* **26**, 73, 1973.

47. J.K.A. Amuzu, B.J. Briscoe and D. Tabor, *Trans. Amer. Soc. Lub. Eng.* **20**(2), 152, 1977.

48. B.J. Briscoe and P.S. Thomas, *STLE Tribology Trans.* **38**(2), 382, 1995.

49. G.A.D. Briggs and B.J. Briscoe, *Nature*, **260**, 5549, 381, 1976.

50. K.C. Ludema and D. Tabor, *Wear*, **9**, 329, 1966.

51. Y. Uchiyama, *Tribology Int.* (in press), 1998.

52. B.J. Briscoe and S.L. Kremnitzer, *J. Phys. D. Appl. Phys.* **12**, 505, 1979.

53. M.J. Adams and B.J. Briscoe, Granular Matter: An Interdisciplinary Approach, ed. A. Mathias, Springer-Verlag, New York, 259, 1993.

54. B.J. Briscoe, *Chem. Eng. Sci.* **42**, 713, 1987.

55. B.J. Briscoe, G.S. Davies and T.A. Stolarski, *Tribology Int.* **17**, 129, 1984.

56. B.J. Briscoe and T.A. Stolarski, *Wear*, **112**, 371, 1986.

57. B.J. Briscoe and A.C. Smith, *J. Appl. Poly. Sci.* **28**, 3827, 1983.

58. B.J. Briscoe and P.J. Tweedale, Tribology of Composite Materials, *Proc. Am. Soc. Mat. Conf.*, Oak Ridge, Tennessee 1990.

59. B.J. Briscoe, M.J. Adams and T.K. Wee, *J. Phys. D. Appl. Phys.* **23**, 406, 1990.

60. B.J. Briscoe, M.J. Adams and A. Arvanitaki, *Proc. 23rd Leeds-Lyon Symp.* 1996, Elsevier, Holland, 1996.

61. B.J. Briscoe and F. Motamedi, *Wear*, **158**, 229, 1992.

62. M.J. Adams, U. Tuzun and B.J. Briscoe, *Chem. Eng. Sci.* **43**, 1083, 1988.

63. B.J. Briscoe, M.J. Adams and L. Pope, *Powder Technology*, **37**, 169, 1984.

64. B.J. Briscoe, M.J. Adams and F. Motamedi, *J. Phys. D. Appl. Phys.* **26**, 73, 1993.

65. H.A. Janssen, *Z. Ver. Deutsch. Ing.* **39**, 1045, 1895.

66. B.J. Briscoe and N. Ozkan, *Powder Technology*, **90**, 195, 1997.

67. B.J. Briscoe and N. Ozkan, *Ceramics Int.* **23**, 521, 1997.

68. B.J. Briscoe and P.D. Evans, *Power Technology*, **65**(1–3), 7, 1991.

69. B.J. Briscoe, I. Aydin and N. Ozkan, *Int. Mach. Tools Manufacture*, **35**(2), 345, 1994.

70. B.J. Briscoe and S.L. Rough, *Colloids and Surfaces* (in press).

71. B.J. Briscoe, A.C. Smith and M. Fernando, *J. Phys. D. Appl. Phys.* **18**, 1075, 1985.

72. B.J. Briscoe and M.J. Adams, Part. C, Food and Bioproducts, *Trans. I. Chem. E.* **71**, 251, 1994.

73. X. Weert, PhD Dissertation, Extrusion of Powders, Imperial College, 1998.

74. B.J. Briscoe, I. Aydin and K.Y. Sanliturk, *Comp. Mat. Sci.* **3**, 28, 1994.

75. B.J. Briscoe, I. Aydin and N. Ozkan, *MRS Bull.* **22**, 45, 1997.

76. B.J. Briscoe, I. Aydin and K.Y. Sanliturk, *J. Eur. Cer. Soc.* **17**, 1201, 1997.

77. B.J. Briscoe and N. Ozkan, *Ceramics Int.* **23**, 521, 1997.

78. B.J. Briscoe, A.U. Khan, P.F. Luckham and N. Ozkan, *5th European Ceramic Conf.*, Versailles, France, June 1997.

79. B.J. Briscoe and N. Ozkan, to be published in *J. Mats. Res.* 1997.

80. B.J. Briscoe, M.J. Adams, E. Pelillo and S.K. Sinha, Third Body Phenomena, *Proc. Leeds/Lyon Meeting, 1995,* ed. D. Dowson *et at.*, Elsevier, 535, 1996.

81. B.J. Briscoe, M.J. Adams and M. Kamyab, *Adv. Colloids Interface Sci.* **44**, 143, 1993.

82. B.J. Briscoe, M.J. Adams, S.K. Biswas and M. Kamyab, Advances in Particle Technology, Special Issue, *Powder Technology*, **65**(1–3, 381–392), 1991.

83. B.J. Briscoe and others, MAFF/DTI Report, *Solid Process Engineering*, MAFF/DTI Issue 1998.

84. B.J. Briscoe, M.J. Adams, I. Aydin and S.K. Sinha, *J. Non-Newtonian Fluid Mech.* **71**, 41, 1997.

85. B.J. Briscoe, M.J. Adams and S. Shamasundar, *Computational Plasticity, Fundamentals and Applications*, Part II, eds. Dr. J. Owen, E. Onate and E. Hilton), Pineridge Press, Swansea, 1992.

86. J. Benbow and J. Bridgwater, Paste Flow and Extrusion, Oxford Series on Advanced Manufacturing, Vol. 10, Clarendon Press, August 1993.

87. B.J. Briscoe, G.M. Corfield, C.J. Lawrence and M.J. Adams, *Trans. I. Chem. E.* Part A, **76**, January 1998.

88. B.J. Briscoe, M.J. Adams, X. Weert and G. Corfield, *Chem. Eng. Sci.* (in press), 1998.

89. J.A. Greenwood and J.B.P. Williamson, *Proc. Roy. Soc. London*, **A295**, 300, 1966.

90. B.J. Briscoe and S.S. Panesar, *J. Phys. D. Appl. Phys.* **19**, 841, 1986.

91. B.J. Briscoe and S.S. Panesar, *J. Adhesion Sci. Technol.* **2**(4), 123, 1988.

92. B.J. Briscoe and S.S. Panesar, *Proc. Roy. Soc. London*, **A433**, 23, 1991.

93. B.J. Briscoe, M.B. Khan and S.M. Richardson, *Plastics Eng.* 43–45, November 1988.

94. B.J. Briscoe, M.B. Khan and S.M. Richardson, *Plastics and Rubber Processing and Applications*, **10**, 65, 1988.

95. B.J. Briscoe, E. Pelillo, F. Ragazzi and S.K. Sinha, to be published in *Polymer*, 1997.

96. B.J. Briscoe and N. Ozkan, *J. European Ceramics Soc.* **17**, 1675, 1997.

97. B.J. Briscoe and S. Sebatian K., *Proc. Roy. Soc. London*, **A452**, 439, 1996.

98. B.J. Briscoe, D. Liatsis and D. Gritsis, *Phil. Trans. Roy. Soc. London*, **A339**, 498, 1992.

99. B.J. Briscoe and S. Zakaria, *J. Mat. Sci.* **25**(6), 3017, 1990.

100. B.J. Briscoe and C.T. Kelly, *E-MRS Symposium*, *Strasbourg*, November 1992. Also published in *Mat. Sci. Eng.* **A168**, 111, 1993.

101. B.J. Briscoe, T. Savvas and T.C. Kelly, *Rubber Chem. Tech.* **67/3**, 384, 1994.

102. B.J. Briscoe and J. Klein, *Nature*, **266**, 5597, 1977.

103. B.J. Briscoe and J. Klein, *Proc. Roy. Soc. London*, **A365**, 53, 1979.

104. B.J. Briscoe, B.H. Stuart and S. Rostami, *Spectochemica Acta*, **A49**(5–6), 153, 1993.

105. B.J. Briscoe, B.H. Stuart, P.S. Thomas and D.R. Williams, *Spectrochemical Acta*, **A47**, 1299, 1991.

106. B.J. Briscoe, B.H. Stuart, S. Sebastian and P.J. Tweedale, *Wear 1992 ASME Mtg.*, San Francisco, Wear Special Issue, *Wear*, **162–164**, 407, 1993.

107. B.J. Briscoe and K. Galvin, *Colloids and Surfaces*, **52**, 219, 1991.

108. B.J. Briscoe and K. Galvin, *J. Solar Energy*, **46**(4), 191, 1991.

109. B.J. Briscoe and K. Galvin, *Phys. Rev.* **A43**(4), 1906, 1991.

110. B.J. Briscoe, M. Pickles, K. Julian and M.J. Adams, *Wear*, **181–183**, 1995.

111. B.J. Briscoe, P.F. Luckham and S.R. Ren, *Phil. Trans. Roy. Soc. London*, **A348**, 179, 1994.

112. B.J. Briscoe, P. Luckham and S. Zhu, *Macromolecules*, **29**(19), 6208, 1996.

113. B.J. Briscoe and R. Nosker, *Wear*, **95**, 214, 1984.

114. B.J. Briscoe and R. Nosker, *Polymer Comm.* **26**, 307, 1985.

Professor A.J. Kinloch
DSc (Eng), FREng, FIM, FRSC

Professor Kinloch was born in 1946 in London. He received his first degree in Polymer Science and Technology, and then in 1972 a PhD from Queen Mary College, University of London, for his research on the mechanics and mechanisms of adhesion. He then started work for the Ministry of Defence, and undertook research and development in the area of adhesion, adhesives and polymer science. In 1984, he joined the Department of Mechanical Engineering, Imperial College, as "Reader in Engineering Adhesives", and in 1990 was appointed "Professor of Adhesion". He has published over one hundred and fifty patents and papers in the areas of adhesion and adhesives, toughened polymers and the fracture of polymers and fibre-composites; and written and edited seven books in these areas. He has served on many EPSRC, Institute of Materials, Institution of Mechanical Engineering and Government Committees, and he has been a Visiting Professor at Universities in Europe and the USA. In 1988, he was awarded a DSc (Eng) from the University of London for his research in Materials Science. In 1992 he received the US Adhesion Society Award for "Excellence in Adhesion Science" and was elected a "R.L. Patrick Fellow" in 1995. Also, in 1994 he received the Adhesion Society of Japan Award for "Distinguished Contributions to the Development of Adhesion Science and Technology". In 1996, he was awarded the "Griffith Medal and Prize" from the Institute of Materials for his work on the application of fracture mechanics to studies of adhesion and adhesive. In 1997, he was elected to the Royal Academy of Engineering.

STICKING UP FOR ADHESIVES

A.J. KINLOCH, FREng

University of London
Imperial College of Science, Technology and Medicine
Department of Mechanical Engineering
Exhibition Road
London, SW7 2BX, UK

1. Introduction

The problem with giving an Inaugural Lecture on the subject of "Adhesion and Adhesives" is that everyone is familiar with the subject, since everyone uses "glues". Now some people are "true believers" and will try using adhesives to join anything to anything, as is shown in a video sequence of television advertisements for "Solvite" wallpaper-paste adhesive. These advertisements show "Solvite" adhesives being used to stick wallpaper to a steam train-engine; to stick a man in a flying-suit to a wooden board, and then suspending him over the city of Miami under a helicopter; etc. Obviously, some people have complete faith in the power of adhesives!

On the other hand, many people have no faith in adhesives at all, and such "heretics" therefore take a "belt and braces" approach when using adhesives. For example, Lady Macbeth clearly believed in always using additional mechanical fastening methods to ensure success, since she stated:

"We fail!
But screw your courage to the sticking-place
And we'll not fail."

(Lady Macbeth to Lord Macbeth discussing the murder of Duncan; Act 1, Scene 7.)

However, I do hope that this Inaugural Lecture will convince you that by "Sticking Up for Adhesives" that we have much to gain.

The transport industry provides some excellent examples of the use of adhesives in critical and demanding engineering applications. Indeed, the use of adhesives in aerospace goes back a very long way, but not always with complete success. Since, one of the earliest flying devices to use adhesives was the "flying-wing structure" designed and built by Daedalus, possibly the engineer of the family. This structure was flown, of course, by his son, Icarus. He was possibly the chemist in the family who selected the beeswax adhesive but clearly neglected to study the heat stability of this particular type of glue — a tragic oversight that cost him his life.

This introduces another theme in the story of "Adhesion and Adhesives". Namely, that it is a truly multidisciplined subject and requires the skills of the engineer and chemist, as well as the physicist, in order to achieve innovative and successful results.

Fortunately, more recent flying machines have used adhesives with far greater success. The early airships and biplanes of this century used adhesives based upon casein, which is a natural material and is a by-product of milk. These adhesives worked well, except when they got wet. They then became very weak and smelt of old camembert cheese. However, engineers are a cunning breed, since it is claimed that they used this fact as an early form of nondestructive test. The aircraft engineers routinely smelt the bonded parts of the aircraft, and when the joints smelt of old camembert cheese they knew that the adhesive joints were about to fall apart, and that the adhesive should be replaced.

The problem of the poor ageing of adhesives based upon natural materials was largely overcome by the introduction of synthetic, polymeric adhesives. For example, in the Second World War the very successful "Mosquito" aircraft relied mainly upon urea-formaldehyde resin adhesives to bond together its wooden structure. This type of adhesive was, however, rather brittle in nature, and was prone to cracking and fracturing. Other types of early synthetic adhesives more suitable for bonding metals together, such as the phenolic resins, were also rather brittle materials. Thus, a major development in the 1940s was the modification of the chemistry of such very brittle, thermosetting resins to give tougher adhesives, and a very important type was based upon a combination of vinyl-formal/phenolic-resin polymers. This invention represented a major development in adhesives technology and enabled metallic, as well as wooden, components to be bonded very successfully. For example, vinyl-formal/phenolic-resin adhesives were extensively employed by the designers of the "Comet" jet airliners of the 1950s, particularly to give both high stiffness and strength, coupled with a relatively low weight, to the all-metal fuselage and wings.

These pioneering developments in the 1940s and 1950s by British scientists and engineers has led to the construction of modern aircraft being dependent upon the use of adhesives. The engineering adhesives used today are all based upon synthetic polymers, such as modified-phenolic, epoxy, acrylic and urethane polymers. They are employed, for example, to attach stiffening stringers to the fuselage- and wing-skins, as shown in Figs. 1 and 2, and to enable honeycomb structures to be made, as shown in Fig. 3. In these applications, the use of adhesives allows stiff and strong, but lightweight, components to be manufactured. It should be noted that in Fig. 2, the two gentlemen are undertaking the modern version of the "old camembert cheese"

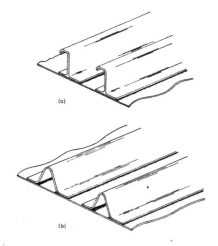

Fig. 1. Schematic showing stiffening stringers which are adhesively-bonded to skin panels.
(a) Extruded *J*-section stringers.
(b) Rolled-strip closed-channel stiffener.

Fig. 2. In the "British Aerospace 146" one of the largest components is the wing-skin assembly which is manufactured using aluminium alloy. The stiffening stringers are bonded onto the skin using a modified-phenolic adhesive. The two gentlemen are conducting nondestructive tests on the bonded joints.

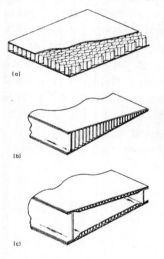

Fig. 3. Schematic showing honeycomb structures which are manufactured by adhesively-bonding skins to a honeycomb core.
(a) Honeycomb panel (flooring, etc.).
(b) Honeycomb structures for control surfaces, trailing edges, etc.
(c) Honeycomb structures for aerofoils.

test. However, their modern nondestructive test method uses ultrasonics, rather than their noses, to detect whether the adhesive is satisfactory or not.

Helicopters also rely upon adhesives and Fig. 4 shows a section of a helicopter blade. This blade uses several sections of stainless steel for the leading edge and a honeycomb trailing-edge (based upon a plastic impregnated core and glass-fibre reinforced plastic skins), and all these different materials are joined together using adhesives. Although, whether the pilots are actually told that they are held up by "glue" is doubtful!

Fig. 4. A section of the blade for the Westlands "Lynx" helicopter. The various materials used are stainless steel, plastic-impregnated paper honeycomb-core and glass-fibre reinforced-plastic skins. These various materials are adhesively-bonded to form the blade.

Fig. 5. The upper shaft is a carbon-fibre reinforced-plastic (CFRP) drive-shaft, with adhesively-bonded metal end-fittings, for the "Peugeot" rally car. The lower tube is manufactured from "Kevlar"-fibre reinforced-plastic, again with an adhesively-bonded metal end-fitting. The lower tube is a static tube in which the drive shaft rotates, so as to protect the CFRP shaft from impact damage.

Switching modes of transport, then the "Peugeot" rally car has a carbon-fibre reinforced-plastic drive shaft which the metal end-fittings bonded onto the drive shaft, as illustrated in Fig. 5. The adhesive, a toughened-epoxy polymer, has therefore to withstand the very high torque from the powerful engine of this rally car. Even faster are the latest designs of sports cars which are made from aluminium alloy to keep the weight of the vehicle relatively low. Such alloys cannot be readily spot-welded, and therefore adhesive bonding is employed as the joining method. The use of lightweight materials, such as aluminium-alloys, plastics and fibre-composites, also leads to excellent fuel economy. Hence, the adhesive-bonding of these types of materials to manufacture cars, lorries, buses, trains, etc. is a yet another rapidly developing area for the adhesives engineer.

Going on to two wheels, the US Olympic team recently used a racing bicycle constructed from lightweight aluminium-magnesium alloys with the various components being adhesively-bonded, again with a toughened-epoxy adhesive being employed. Adhesives were also used in the manufacture of the bike ridden to victory by Chris Boardman in the last Olympic Games. His bike was made from carbon-fibre reinforced-plastic (CFRP) components which were glued together. Whilst on the subject of bicycles, a postgraduate student from a joint engineering-design course organised by Imperial College and the Royal College of Art recently designed and made a lightweight folding bike from aluminium alloy and plastic components. Needless to say, the various components were joined together using adhesives. Furthermore, at least one willing "volunteer" from the audience had actually sufficient faith in adhesives to ride this bike around the lecture theatre. (And a successful bicycle ride by Dr. B.R.K. Blackman, of Imperial College, was observed to take place.)

Now in all these many different applications of adhesives in engineering structures it is possible to identify three distinct stages in the formation of an adhesive joint. First, the adhesive initially has to be in a "liquid" form so that it can readily spread over and make intimate molecular contact with the substrates; the substrates being the materials we wish to join. Secondly, in order for the joint to bear the loads which will be applied to it during its service life, the "liquid" adhesive must now harden. In the case of adhesives used in engineering applications, the adhesive is initially in the form of a "liquid" monomer which polymerises to give a high molecular-weight polymer. Thirdly, as engineers we must appreciate that the load-carrying ability of the joint, and how long it will actually last, are affected by: (a) the design of the joint, (b) the manner in which we apply loads to it, and (c) the environment which the joint encounters during its service life.

To understand the science involved, and to succeed in developing the technology, we therefore require the skills and knowledge from many different disciplines. Indeed, we need the input from surface chemists, polymer chemists and physicists, and from design and materials engineers. Thus, the science and technology of adhesion and adhesives is a truly multidisciplined subject.

We have attempted to bring these different disciplines together by developing a "fracture mechanics" approach to the failure of adhesive joints. The concepts of "fracture mechanics" were introduced by A.A. Griffith in the 1920s whilst working at the Royal Aircraft Establishment, Farnborough. He recognised the importance of flaws in a material or structure. These flaws may be molecular-sized inhomogeneities, or air bubbles, or particles of dirt or dust, or they may be actual cracks. However they arise, Griffith proposed that the strength of a material, or structure, is governed by their presence. He proceeded to define a term, the fracture energy, which is the energy needed to propagate a flaw through unit area of the material, or structure. The fracture energy is given the symbol "G_c" — where "G" is for Griffith and the subscript "c" indicates that it is the critical value for crack growth. The importance of flaws may be shown by loading, in polarised light, a material whose refractive index changes with load. If the strip of photoelastic material contains an edge crack, then the concentration of strain and stress around the crack is clearly seen by the intense pattern of colours which develop around the crack tip. I then demonstrated how the ideas of fracture mechanics build upon this fact by suspending a large sheet of paper from a horizontal support, with a dead load of about 300 Newtons applied to the bottom edge of the paper sheet. I then used a sharp knife to make a small cut in one edge of the sheet of paper, and no adverse effect was seen. I then continued to cut deeper into the sheet, so making the edge-crack progressively longer. Nothing was observed to occur, until, at a critical length, the crack propagated extremely rapidly across the width of the loaded paper-sheet. The ideas proposed by Griffith allow us to calculate the size of the crack at which the rapid, catastrophic failure of the paper sheet occurs, and also to deduce the value of the fracture energy, G_c, needed for failure.

Now, we have been developing methods of fracture mechanics with respect to the failure of adhesive joints, so that we can determine the value of G_c for either a cohesive failure of the adhesive, or for an interfacial failure along the adhesive/substrate interface. The advances we have made in the fracture mechanics of adhesive joints have enabled us to understand better the science of and technology of adhesion and adhesives, as I will attempt to show during this Inaugural Lecture.

2. Interfacial Contact and Intrinsic Adhesion

2.1. Introduction

As I mentioned previously, the first stage in the formation of an adhesive joint is concerned with attaining intimate interfacial contact between the adhesive and substrates, and then establishing strong and stable intrinsic adhesion forces across the adhesive/substrate interfaces. In order to achieve these requirements the substrates often have to be subjected to some form of surface treatment before the adhesive is applied.

Now, all these aspects emphasise the importance of surfaces and surface chemistry in the use of adhesives. However, surface science is a fiendishly difficult area of research, and the problems of understanding surfaces were summarised by W. Pauli by the following comment:

"God created solids,
But surfaces are the work of the Devil!"

Notwithstanding, we have persevered in our research on surfaces, and we have used various experimental methods to understand the Devil's work. For example, we have used:

(a) contact angles (i.e. the tangent angle at the contact point of a liquid droplet resting on a solid substrate surface) to determine the surface tension (or surface free energy) of the adhesives and substrates;

(b) X-ray photoelectron spectroscopy to identify the chemical nature of the surface;

(c) ellipsometry to measure the thickness of thin adsorbed adhesive or primer layers;

(d) reflection high-energy electron diffraction to gain information concerning the orientation of such layers;

(e) optical and electron microscopy to determine the surface morphology and topography; and

(f) secondary-ion mass spectroscopy to detect the type of interfacial bonding which is present.

One problem for a university researcher is that to use most of these techniques requires a considerable expenditure, both in terms of capital equipment and running costs. It would be far less expensive for Imperial College if we changed our research areas from subjects such as adhesion science, physics, etc. to philosophy — at least according to Isaac Asimov who once quoted an American University President as saying:

"Why is it that you physicists always require so much expensive equipment? Now the Department of Mathematics requires nothing but money for paper, pencils and erasers ... and the Department of Philosophy is better still. It doesn't even ask for erasers."

2.2. Mechanisms of intrinsic adhesion

However, expensive techniques such as X-ray photoelectron spectroscopy and secondary-ion mass spectroscopy have enabled us to determine exactly why materials do adhere. Many people, for many years, did believe that adhesion between the adhesive and substrates was due to some form of "microscopic" mechanical interlocking. This theory essentially proposes that mechanical keying, or interlocking, of the adhesive into irregularities of the substrate surface is the major source of intrinsic adhesion. One example where mechanical interlocking is of prime importance is in the use of mercury amalgam for filling tooth cavities. The dentist drills out the tooth material to give a relatively large "ink-bottle" pit, ideally with an undercut angle of about 5°, and a mercury-amalgam filling-material is then forced into this cavity. The main mechanism of adhesion which then occurs at the filling/tooth interface is mechanical interlocking. However, the attainment of good adhesion between smooth surfaces, such as adhesives to glass or to mica, exposes the mechanical interlocking theory as not being of general and wide applicability.

Since the intrinsic adhesion between the adhesive and substrates does not typically arise from mechanical interlocking occurring across the interfaces, then how does it arise? The answer to this question is that the adhesion arises from the fact that all materials have forces of attraction acting between their atoms and molecules, and a direct measure of these interatomic and intermolecular forces is "surface tension". The tension in surface layers is the result of the attraction of the bulk material for the surface layer, and this attraction tends to reduce the number of molecules in the surface region resulting in an increase in intermolecular distance. This increase requires work to be done, and returns work to the system upon a return to a normal configuration. This explains why "surface tension" exists and why there is a "surface free energy".

One well-known effect of surface tension acting in water is that it enables insects to walk upon its surface. Another effect, which may be readily demonstrated, is that it enables a steel needle to be supported on the surface of water. However, when the surface tension is lowered, by the addition of a small amount of detergent, which decreases the forces of intermolecular attraction, the weight of the needle can no longer be supported by the surface tension, and the needle sinks.

Finally, it must be appreciated that solids, as well as liquids, possess a "surface tension"; in the case of solids this property is generally termed the "surface free energy". Of course, the effects of a surface tension being present are far less readily observed in solids, than in liquids.

Now, the forces of interatomic and intermolecular attraction may not only act in the bulk and surface layers of liquids and solids, but may also act across the interfaces between phases. Indeed, it is the presence of such forces of attraction which is generally responsible for the intrinsic adhesion between the adhesive and the substrates, and this most basic mechanism of adhesion was recognised by Michael Faraday over one hundred years ago! Thus, we can state that, provided sufficiently intimate molecular contact is achieved at the interface, materials will adhere because of the interatomic and intermolecular forces which are established between the atoms and molecules in the surfaces of the adhesive and substrates. The most common of such forces are van der Waals forces and these are referred to as "secondary bonds". Also in this category may be included hydrogen bonds. In addition, chemical bonds may sometimes be formed across the interface. This is termed chemisorption and involves ionic, covalent or metallic interfacial bonds being established; these types of bonds are referred to as "primary bonds". The terms primary and secondary are in a sense a measure, albeit somewhat arbitrary, of the relative strengths of the interatomic and intermolecular bonds.

To illustrate some of the above aspects I will use two examples taken from our current research work at Imperial College. The first is concerned with the bonding of fibre-composite materials and the second is the development of organometallic primer layers.

2.3. *The bonding of fibre-composite materials*

As was shown earlier, the use of fibre-composite materials which are based upon continuous glass- or carbon-fibres embedded in a polymeric matrix is steadily increasing in many engineering applications. Further, a recent development has been the use of a <u>thermoplastic</u> polymeric matrix. For example, matrices such as poly(ether-ether ketone) or poly(aromatic amides) have been developed and employed, as opposed to the more common thermosetting polymeric matrices based upon an epoxy or an unsaturated-polyester resin. The advantages that the newer thermoplastic matrices can offer include shorter production times, higher toughness and easier recycling.

However, we discovered that the fibre-composites based upon the thermoplastic matrices were difficult to join using conventional epoxy or acrylic engineering-adhesives. Nevertheless, we found that subjecting the composite materials to a "corona-discharge" treatment prior to adhesive bonding was a very effective method of obtaining good adhesion, and high joint strengths. The "corona-discharge" treatment of fibre-composite specimens is shown in Fig. 6. This treatment basically involves applying a high voltage (15–20 kV at a frequency of 15 to 20 kHz) across the air-gap

Fig. 6. "Corona-discharge" treatment of carbon-fibre reinforced-plastic (CFRP) specimens, where the CFRP material is based upon a thermoplastic matrix. This pretreatment allows the fibre-composite materials to be successfully bonded.

between the composite surface and an electrode. The voltage is increased until it exceeds the threshold value for electrical breakdown of the air-gap, when the air is ionised. Hence a plasma, at atmospheric pressure, of excited oxygen, ozone, etc. ions and radicals is generated. These very active ions and radicals then react with the surface layers of the fibre-composite material, and so chemically modify its surface. The modified surface possesses a higher surface free energy; and this leads to better spreading of the epoxy adhesive over the surface of the fibre-composite and to higher intrinsic adhesion forces being established across the adhesive/fibre-composite interfaces. These aspects are reflected in tougher and stronger adhesive joints being made when the fibre-composite is "corona-discharge" treated prior to bonding, as was demonstrated in the Inaugural Lecture.

To understand and quantify the effects of the "corona-discharge" treatment we have used many of the experimental techniques which I listed previously. For example, the surface analytical method of X-ray photoelectron spectroscopy may be used to detect the changes in the surface regions of the fibre-composite before, and after, surface treatment. The technique of X-ray photoelectron spectroscopy is based upon placing the specimen in a ultra-high vacuum chamber and firing X-rays at the surface, but analysing the energies of the photoelectrons which are emitted. Since the photoelectrons can only escape from about the first 3 nm (1 nm = 10^{-9} m), then this method gives information about only the outermost surface regions of the material. X-ray photoelectron spectra of a carbon-fibre poly(aromatic amide) composite, before and after "corona-discharge" treatment, are shown in Figs. 7(a) and 7(b) respectively. The increases in the concentration, and the type of, oxygen-containing chemical groups in the surface regions after subjecting the substrate to a "corona-discharge" treatment may be clearly seen. Further, contact-angle measurements showed that the epoxy adhesive would not spread over the untreated fibre-composites, but after subjecting the fibre-composite to a "corona-discharge" treatment the adhesive did spread readily over the composite's surface, and exhibited a very low contact angle. Thus, the increased presence of oxygen-containing chemical groups due to the "corona-discharge" treatment, as shown by using X-ray photoelectron spectroscopy, has increased the polarity of the surface of the composite and led to an increase in the surface free energy for the treated fibre-composite; and hence we obtain better spreading and intrinsic adhesion of the epoxy adhesive.

Now, it is difficult to demonstrate readily these aspects using the fibre-composites, which are black in colour. However, they can be illustrated using poly(tetrafluoroethylene) (PTFE) as a substrate material, which is white in colour. PTFE is better known by its trade names of "Teflon" and "Fluon", and is of course the material used for the "nonstick" coating of fryingpans and saucepans. The reason that PTFE is "nonstick", and the reason why the thermoplastic composites are also difficult to bond, is that

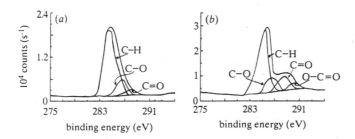

Fig. 7. X-ray photoelectron spectra of the carbon 1s peak of a unidirectional carbon-fibre/poly(aromatic amide) composite.
(a) Untreated.
(b) "Corona-discharge" treated (5 J/mm^2 of energy applied).

all these materials possess very weak surface forces and, therefore, possess very low surface free energies. I can demonstrate this by trying to spread blue ink over the surface of the white PTFE material. The ink does not spread over the surface of the PTFE to give a continuous film of ink, but remains as discrete droplets, which have a high contact angle. Even if I try to make the ink spread, by forcing it over the surface using a spatula, the ink still remains as discrete droplets. Now I can treat the PTFE by placing it in a solution of sodium naphthalenide for about fifteen seconds. This treatment produces a dull brown-black surface layer, which has a far greater surface free energy than the untreated PTFE. These effects arise from the fact that the sodium naphthalenide treatment defluorinates the surface layer of the PTFE and introduces relatively polar, oxygen-containing, groups into the surface regions. (Also, it creates carbon–carbon conjugated double bonds in the surface regions, and it is these groups which cause the brown-black colouration.) The ink-wetting experiment may be repeated on the treated PTFE material, and the ink now readily spreads over the surface of the treated PTFE to give a smooth, continuous film of ink, which exhibits a very low contact angle. The greater adhesion of the treated PTFE may be shown by gluing together two PTFE strips, to form an overlap joint, using a cyanoacrylate (i.e. "Superglue") adhesive. The joints were made and then left for a few minutes for the adhesive to polymerise, and so harden. It was demonstrated that the joint made using the untreated PTFE strips was very weak, and could easily be pulled apart. On the other hand, the joint made using the treated PTFE strips was sufficiently strong to resist all my attempts to break the joint.

To summarise, the reason why surface treatments such as plasma treatments (e.g. the "corona-discharge" method) or chemical-etch treatments (e.g. the sodium naphthalenide solution) are effective is that they introduce chemical groups which are relatively polar, into the surface regions of the substrate material. Hence, stronger intermolecular forces act in the surface regions of the materials, and this leads to an increase in the surface free energy of the substrate. Thus, we can attain (a) better interfacial contact between the adhesive and substrate, and (b) higher intrinsic adhesion forces of molecular attraction acting across the interface.

2.4. *The use of organometallic primers*

A second example of the importance of surface chemistry in adhesive bonding is the use of organometallic primers. The most common type of such primers are the silane-based primers. The types available commercially have the general structure $X_3Si(CH_2)_nY$; where $n = 0$ to 3, X is a hydrolysable group on silicon and Y is a organofunctional group usually selected to be chemical reactive with a given adhesive. The generally accepted mechanism of intrinsic adhesion for such primers is that they enable the formation of strong, primary, interfacial bonds across the substrate/primer/adhesive interfaces. They therefore effectively enable the adhesive to be chemically reacted, and so strongly bonded, to the surface of the substrate. This gives rise to strong intrinsic adhesion, which is reflected in strong and durable (i.e. water-resistant) adhesive joints.

Now we have been developing such primers which form only a "monolayer" in a joint. As the name suggests, a "monolayer" is where only one molecular layer of a chemical species adsorbs onto a surface, and for many years monolayers have been known to be powerful agents for modifying surfaces. We can see what a great influence monolayers may have from an experiment where I ignite a dish containing a saturated aqueous solution of ether and then extinguish the flames by using oleic acid (i.e. $CH_3(CH_2)_7CH = CH(CH_2)_7COOH$). The oleic acid adsorbs as a monolayer via its highly-polar acid-end group. Since oleic acid does not burn, it smothers and so extinguishes the flames. (The experiment, fortunately, worked as planned — thus the fire-brigade were <u>not</u> called upon!) As could be clearly observed, the amount of oleic acid needed to extinguish the flames was very, very small. This is because only one molecular layer was needed and so the thickness of the oleic-acid layer was about 3 nm; and when considering the thickness of this monolayer bear in mind that 1 nm $= 10^{-9}$ m $=$ one millionth of a millimetre.

Now to achieve a monolayer of an organometallic primer adsorbed onto the surface of a substrate, simply from dipping the substrate into a solution of the primer, we have synthesised a long alkyl-chain based silane, namely 18-nonadecenyltrichlorosilane, which has the chemical formulae:

$$CH_2 = CH(CH_2)_{17}SiCl_3$$

The substrate, an aluminium-alloy, was then immersed in a 0.1 Molar solution of this silane primer in 90% hexadecane/10% chloroform. The silane first hydrolyses, to give 18-nonadecenyltrihydroxysilane, and then chemically adsorbs onto the surface of the aluminium alloy, with the hydroxyl groups reacting with similar groups which are present on the aluminium oxide to give a -Si-O-Al- chemical bonds. The presence of the long alkyl chain forces the hydrocarbon chains to pack tightly together and, being nonpolar, they orientate themselves away from the substrate. Thus, an orientated monolayer of the silane primer, about 3 nm thick, and which is chemically bonded to the aluminium oxide, is formed on the surface of the aluminium alloy. The second step is to convert the vinylic-end groups on the adsorbed primer to groups which may react with the adhesive which is to be used to bond the silane-primed aluminium-alloy substrates together. In the present experiments the vinylic-end groups were converted to hydroxyl groups. (It should be noted that, if the vinylic-end groups were converted to hydroxyl groups <u>before</u> the primer was adsorbed onto aluminium alloy, the presence of polar groups at both ends of the primer molecule would have prevented the formation of an orientated monolayer of the silane primer.) A polyurethane adhesive was then used to join the silane-primed aluminium-alloy substrates.

The measured fracture stress of the joint, as a function of the length of the sharp crack which has been placed at the interface in the joint, was then determined and the results are shown in Fig. 8. As may be seen, when the fracture stress is plotted as a function of the inverse of the square-root of the crack length, an excellent linear fit is found for any given type of joint, as demanded by the theory of fracture mechanics developed by Griffith. For the joints where the vinylic-end group on the primer was not changed, so giving a primer which would <u>not</u> react with the incoming adhesive, the adhesive fracture energy, G_c, is lower than that of the control joints; where no primer was employed. However, when the vinylic-end group on the primer was changed, so as to give a primer which would now react with the incoming adhesive, the joints possessed an adhesive fracture energy, G_c, which was significantly greater than that of the control, unprimed joints.

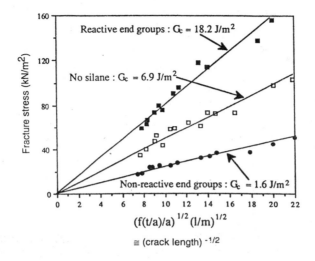

Fig. 8. The effect of using a silane-based primer in polyurethane adhesive/aluminium-alloy adhesive joints. The figure shows a plot of the fracture stress versus a function of the inverse of the square root of the length of the crack inserted at the interface. The value of the adhesive fracture energy, G_c, may be obtained from the slope of the linear relationship for a given joint. (The control joints employed no primer.)

Thus, the presence of an orientated monolayer of the organometallic silane primer (which is therefore an inherently strong interlayer) and which can chemically react with both the aluminium oxide and the adhesive (to form a primary, chemically-bonded interlayer) dramatically increases the toughness of the joint. Further, if the joint is immersed in water, or exposed to a high relative humidity, then the presence of such a primer may also greatly increase the resistance of the interface of the joint to attack by moisture. Thus, the durability and service life of the joint may be significantly increased.

3. Hardening the Adhesive

3.1. *Introduction*

The second stage in adhesive bonding involves "hardening" (sometimes called "curing") the adhesive. This is necessary since, to achieve spreading of the adhesive over the substrate and establish interfacial molecular contact and adhesion, the adhesive has to be in a "liquid" form. However, for the adhesive joint to be able to bear loads, the adhesive must now harden to form a relatively strong and rigid solid.

This brings us to the role of polymer chemistry and physics in the science and technology of adhesion and adhesives, and **we** physicists can now put the rest of the science and engineering community in their rightful place. At least according to Rutherford, who once stated:

"The only true branch of science is physics, the rest is just like collecting postage stamps."

On the other hand, **we** chemists did fight back against this attitude, thanks to Frederick Soddy who said:

"Chemistry has been termed by the physicist as the messy part of physics, but there is no reason why the physicists should be permitted to make a mess of chemistry when they invade it."

3.2. *Chemical aspects*

The importance of hardening the adhesive can be readily demonstrated by trying to bond together two pieces of balsa wood, using water as the glue. Although the water spreads readily over the balsa wood, the strength of the overlap joint between the pieces of wood is obviously low, since the liquid water has no significant <u>bulk</u> strength. If we repeat the experiment, but now freeze the water by dipping the joint in liquid nitrogen for a few seconds, then whilst the water is frozen — so the adhesive is now ice, of course — the joint is relatively strong. Indeed, when I do break the joint it actually fails in the balsa wood (i.e. substrate) away from the bonded region and not in the adhesively-bonded overlapped region.

Now in the case of engineering adhesives we need a hardening method which is not as temperature sensitive, and which gives more durable joints, than "freezing". And most engineering adhesives harden via the formation of a crosslinked, thermosetting polymer. An excellent example of this polymer chemistry in action, and one which many of the audience will have used themselves, is "Two-Tube Araldite" — a product sold by all good "Do-it-Yourself" shops! As I demonstrated, you simply first mix together the resin (which is an epoxy resin) from one tube and the crosslinking agent (quite logically called the "hardener") from the second tube. You then use the adhesive whilst it is in the "liquid" form, so that you can spread it over the substrates. At this stage the joint has virtually no strength, since the adhesive is still a "liquid". However, the adhesive then hardens over the next few hours, which is achieved by the epoxy resin reacting with the hardener to form a crosslinked, thermosetting polymer. Once hardened, you can apply quite high loads to the polymeric adhesive, without the adhesive undergoing plastic deformation (i.e. without the adhesive flowing) or fracturing.

3.3. *Multiphase adhesives*

For adhesives to be used in very demanding engineering applications, the chemistry of the hardening process is often designed so as to give an adhesive which possesses a "multiphase" microstructure. Indeed, the formation of a multiphase microstructure is crucial in enabling adhesives to be used as very tough, engineering materials. Typically, a liquid rubber is dissolved in the "liquid" (i.e. uncured, monomeric adhesive) and as the adhesive polymerises (and so hardens) the rubbery component phase separates and forms small spherical particles of rubber in the adhesive material. The small rubber particles can be clearly seen in a transmission electron micrograph, see Fig. 9, and they are about 1 to 5 micrometers in diameter (1 mm = 10^{-6} m).

Fig. 9. A transmission electron micrograph of a rubber-toughened epoxy adhesive. The spherical rubber particles in the matrix of cured epoxy polymer may be clearly seen.

To demonstrate the effects of toughening the adhesive by the inclusion of such rubber particles I bonded together two strips of aluminium-alloy. For one joint I used a simple, i.e. nontoughened, adhesive, whilst for the other I used the same basic adhesive but which now possessed an internal microstructure of rubber particles, as shown in Fig. 9. In the case of the nontoughened adhesive, the joint broke easily when I bent the lap joint. However, for the toughened adhesive, I was able to bend the lap joint repeatedly backwards and forwards, without the adhesive failing.

Now, these experimental results do clearly show that the rubber-modified, multiphase adhesives are indeed extremely tough. However, I have always tried to follow Sir Arthur Eddington's rule:

"It is a good rule not to put too much confidence in experimental results until they have been confirmed by theory."

Therefore, over the last few years we have developed theoretical models to explain why such adhesives are so very tough.

We have theoretically modelled (a) the microstructure, (b) the toughening mechanisms, and (c) the resulting fracture energy, G_c, using finite-element stress analysis. Figure 10 shows a computer-generated model of two rubber particles, and we only need to model one quarter of each particle because of the symmetry of the structure. Between the rubber particles is the

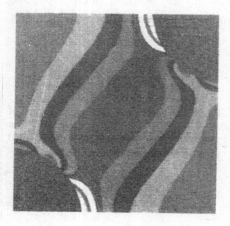

Fig. 10. A computer-generated model of the deformation behaviour of a rubber-toughened epoxy adhesive. The finite-element analysis shows two (one-quarter) spherical rubber particles. The material between the rubber particles is the epoxy adhesive. The green-coloured band represents the development of a plastic shear band in the epoxy polymer — initiated by, and running between, the rubber particles. (The adhesive is being loaded vertically.)

cured adhesive, in this case we have fed into the model the mechanical (i.e. the stress versus strain) properties of a typical crosslinked epoxy polymer. Next we can apply a load to our model, and the colours seen in Fig. 10 represent different levels of stress which result from the interactions of the microstructure and the applied load. The band of colours running at approximately 45° between the rubber particles shows the development of one of the many plastic shear bands which are initiated by the presence of the rubber particles and form in the epoxy adhesive. These plastic shear bands absorb mechanical energy and so reduce the stress concentrations at the tip of any crack, or flaw, in the joint. Hence, the presence of the rubber particles greatly increases the toughness of the adhesive. From the computer model we can theoretically predict the level of toughness of the multiphase adhesive for a given microstructure, provided we measure the stress versus strain behaviour of the rubber and epoxy phases so that we can feed these properties into our model. The values of the fracture energy, G_c, for a

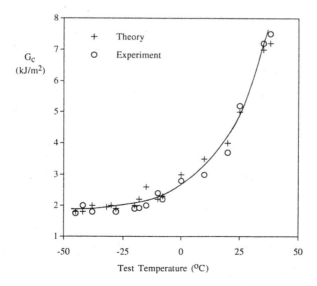

Fig. 11. Values of the fracture energy, G_c, of a rubber-toughened epoxy plotted as a function of the test temperature. Note the good agreement between the theoretical predictions from the modelling studies and the experimental measurements.

rubber-toughened epoxy adhesive from the theoretical modelling studies are compared to the experimentally measured values in Fig. 11. Clearly, following Sir Arthur Eddington's rule, we may now have confidence in the experimental results. Such modelling methods are currently being used in order to develop even tougher adhesives in a more efficient, and less empirical, manner.

The development of these multiphase adhesives represents another major development in adhesives technology, and many industries now rely upon such materials. For example, the "Jaguar XJ220" sports car employs such adhesives extensively for joining the aluminium-alloy parts used in its construction. Indeed, if you could spare about £400,000, then you could help in "Sticking Up for Adhesives" by buying one of these vehicles!

4. Predicting the Strength and Service Life of Adhesive Joints

4.1. *Introduction*

Turning now to the third and last stage, then the design of the joint, the way in which loads are applied to it and the service environment that it must withstand will all affect its mechanical performance and service life. So, now the skills and knowledge of the materials engineer are now demanded.

For example, the combination of hot/wet climatic conditions and cyclic fatigue loads represents a very demanding environment for an adhesively-bonded joint. Indeed, there are cases where engineering components have prematurely failed under such conditions after they have been in-service for periods of just a few months. Obviously **we** engineers wish to be able to predict, and prevent, the failure of adhesively-bonded components. Therefore, accelerated ageing tests are often undertaken during the design phase of an engineering project. Accelerated ageing tests typically involve exposing bonded joints in water, or the environment of interest (e.g. a corrosive salt-spray), at a relatively high temperature. For example, maybe six months exposure in boiling water, or at least water at, say, 60°C. A major problem which may be encountered with such an approach, and a reason why such accelerated tests may be very misleading is succinctly summed up by the question:

"When did boiling an egg ever produce a chicken?"

Thus, it is important to ensure that accelerated tests are selected which do give the same outcome (i.e. the same mechanisms of ageing) as would be seen in real life — the aim being to accelerate the mechanisms of environmental attack, not to produce mechanisms different to those seen in real life.

4.2. *The fatigue behaviour of adhesive joints*

We have been working to predict the lifetime of bonded joints which are subjected to cyclic fatigue loads. Such loading occurs when the joint is subjected to oscillating loads, and this type of loading may be particularly damaging to all types of materials and components. A fracture mechanics test has been used to generate the data shown in Fig. 12. Here the rate of crack growth per cycle, *da/dN*, is plotted against the maximum adhesive

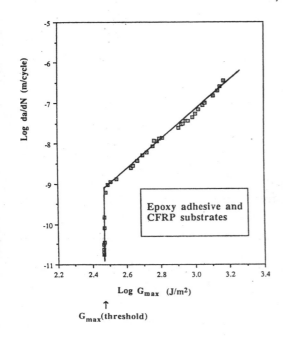

Fig. 12. Results from cyclic fatigue tests conducted on adhesive joints which consist of an epoxy adhesive bonding carbon-fibre reinforced-plastic substrates. The crack growth rate, *da/dN*, per cycle is shown as a function of the maximum fracture energy, G_{max}, in any loading-unloa-ding cycle. Note the presence of a lower bound, threshold, value of G_{max}, below which no cyclic crack growth will occur. (Frequency of testing: 5 Hz.)

fracture energy, G_{max}, which occurs in a loading-unloading cycle. It may be seen that the lower the value of the maximum load in the cycle, and hence the lower the value of G_{max}, then the slower is the rate of crack growth through the joint. Further, there is a threshold value of G_{max}, below which no crack growth occurs. This threshold value obviously provides a very useful design limit when used with appropriate safety factors. Since, providing the maximum load applied to the bonded component is always below the corresponding value of this threshold value of G_{max}, then no failure due to fatigue loading should ever be observed.

Now, these fracture mechanics data shown in Fig. 12 may also be employed to predict the actual fatigue life of other designs of adhesive joints and bonded components. The basic approach is to derive a relationship for the results shown in Fig. 12, and then integrate this relationship. For example, the data may be used to predict the number, N_f, of cycles needed to cause failure of a lap joint, which consists of two strips of fibre-composite material overlapped and bonded together, when subjected to a given cyclic fatigue load. Indeed, we have obtained good agreement between the experimental results and the theoretical predictions for the number, N_f, of cycles such joints may be subjected to before failure is observed. We are currently developing such methods to enable us to predict accurately the lifetimes to be expected from different designs of bonded components.

5. Conclusions

I trust that the examples shown in this Inaugural Lecture have greatly increased your faith in the use of adhesives. But I now come to a final, and extremely convincing, reason for "Sticking Up for Adhesives" — which is that nature, invariably, has anticipated the efforts of mankind. Indeed, nature has most successfully acted as **both** the chemist and the engineer for many, many thousands of years.

An excellent example of nature using adhesion and adhesives is illustrated by the survival method of the plant "Drosera Rotundiflora", otherwise known as the "Sundew Plant". This plant survives by digesting the body fluids of dead ants, but first, of course, it has to capture the ant. Now, to capture the ant it <u>does not</u> drive a nail through the ant, or use mechanical fasteners. It does, however, rely upon "glue". At the end of each stalk on the plant there is a globule of glue. Photographs of the capture of the ant show that the glue spreads readily over the body of the ant, making a low contact angle, and the glue is very tough. As in the case of engineering adhesives, these factors combine to give excellent adhesion and joint strength. However, nature's efforts clearly score over those of mankind, since the "Sundew Plant" does not, of course, have to pretreat the body of the ant before its glue can adhere, not did it rely upon mathematical formulae and modelling to develop its glue.

However, if you are still not sufficiently convinced to become a convert and have faith in adhesives, and so join me in "Sticking Up for Adhesives", then I would suggest that you follow the advice of J.E. Gordon:

"When all else fails
Use bloody great nails."

Acknowledgements

The author would like to acknowledge the assistance of British Aerospace, Ciba Composites and Polymers, Ford Motor Co., Permabond Adhesives, University of Surrey, Westland Helicopters, many colleagues at Imperial College and Mr. W.A. Dukes, formerly of the Ministry of Defence, for their assistance with exhibits, demonstrations, etc. used for the Inaugural Lecture. The present paper appeared in the "Proceedings of the Royal Institution of Great Britain", Edited by P. Day, Volume 67, 1996, and is published with permission of the Oxford University Press.

Bibliography

1. R.H. Bogue, *The Chemistry and Technology of Gelatin and Glue*, McGraw-Hill Book Company, Inc., New York, 1922.
2. N.A. De Bruyne and R. Houwink (Eds.), *Adhesion and Adhesives*, Elsevier, Amsterdam, 1951.

3. Bonded Structures Division, *Bonded Aircraft Structures*, Bonded Structures Division, Duxford, 1957.

4. D.D. Eley (Ed.), *Adhesion*, Oxford University Press, London, 1961.

5. W.C. Wake, *Adhesion and the Formulation of Adhesives*, Applied Science Publishers, London, 1976.

6. A.J. Kinloch, *Adhesion and Adhesives: Science and Technology*, Chapman and Hall, London, 1987.

Professor D.B. Holt

Professor Holt was born in South Africa but grew up in the United States where he had all his school education. Returning to South Africa, he obtained the BSc degree in Physics and Mathematics and the MSc in Solid State Physics at the University of the Witwatersrand in Johannesburg.

He obtained his PhD in Physical Metallurgy at the University of Birmingham in the UK and returned to lecturing and research positions in the Physics Department of the University of the Witwatersrand for five years.

On leaving South Africa he came to the Department of Metallurgy (now Materials) of Imperial College where he has remained ever since apart from a year spent at the University of New South Wales in Sydney, Australia and shorter periods at Bell Laboratories, Murray Hill, New Jersey; MASPEC Institute, Parma, Italy; the Indian Institute of Technology, Kanpur; etc.

MAGICAL MATERIALS FOR MOTIONLESS MACHINES

D.B. HOLT

Department of Materials
Prince Consort Road
London, SW7 2BP, UK
E-mail: d.b.holt@ic.ac.uk

Abstract

Beginning in prehistoric times many materials have been important for their strength and toughness and have been used to make tools and machines to lighten or replace human labour. Since the emergence of modern metallurgy, during the first industrial revolution, when power was being applied to machinery to mass-produce consumer goods, most (at first, nearly all) the emphasis in materials development was on the strength and toughness of materials. This is the better-known half of the story of materials. The other half of the materials story concerns other materials, the precious metals and naturally occurring crystals, that, also from the earliest times, were valued for their transparency, colour, sparkle, shape and their supposed magical properties and were used for decoration and for religious or magical purposes. The physical properties of such crystals and related materials have been exploited in recent decades to make possible modern "solid-state" electronic and optoelectronic systems.

The first report that a semiconducting compound had been synthesised and, surprisingly, found to be transparent, attracted me into research on such solid state materials. We found that this material, GaP, emitted light from grain boundaries. The emission mechanism was established using voltage contrast in a scanning electron microscope for perhaps the first time. This was my entree to research on defect and device structures in micro- and optoelectronic materials and the development of scanning electron micⁿoscope techniques for this field, some aspects of which will be presented.

The control of defects and structure in monolithic semiconductor devices became the enabling technology of the new phase of industrialisation. In this present, second industrial revolution, information technology is spreading throughout the world economy and all aspects of society. The dominant semiconductor and other materials for this technology are important for virtually all their properties other than strength and toughness. The semiconductor devices, for which strength is of virtually no importance, are assembled into machines and systems that do no mechanical work. These machines, like computers, television sets, and telecommunication systems manipulate electrons or photons representing information and increasingly entertain, educate us and order our lives. If computers come to think as well, these materials will have proved magical indeed.

1. Introduction

1.1. *Half the story of materials*

One good way to draw attention to the vital importance of materials in human progress is to remind ourselves of the archaeologists' division of the earliest stages of development, in terms of the most (practically) important materials of these periods, as the stone, bronze and iron ages.

Let us look at the earliest stages a little more closely. I entered the University of the Witwatersrand in Johannesburg, South Africa in 1947, after growing up and having my school education in the United States. I was interested not only in science but also in history. I found that at Wits University there was great interest in prehistory. This was because then-recent local discoveries had extended knowledge of the premodern-human past back to enormously more distant times. Bones of the oldest prehumans,

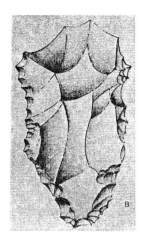

Fig. 1. Acheulean hand axe. Note the beautiful workmanship showing the high level of mastery of the technology of chipping (pressure cleaving) of stone such as flint. [Ref. 3, reproduced, with permission, from the (Encyclopaedia Brittanica 15th Edition Encyclopaedia Brittanica Inc., Chicago) Macropaedia Vol. 8, p. 608.]

the hominid australopithicines, evolutionary "missing links", had been found in the nearby Sterkfontein caves suggesting that Africa was the original cradle of mankind. Archaeologists were then just beginning to guess that some of the crudely chipped stones found lying in large numbers in the same strata might be tools, made by the hominids and vastly earlier than the Paleolithic tools of Europe and elsewhere of the then so-called "Old Stone Age" (the paleolithic era). The latter extended back a half a million years in Europe and was characterised by the Acheulean style of stone artefacts typified by the beautifully chipped Acheulean hand axe (Fig. 1).

The decades of research since then, especially at the even better, later-discovered hominid site at Olduvai Gorge in Kenya, has confirmed that the earliest predecessors of modern mankind did first appear in Africa. It also confirmed that the characteristic handy-sized rocks with a chipped section along one edge, found in these sites were the oldest form of tool, now known as the olduwan chopper-scraper (Fig. 2). These tools, dating back to 2,600,000 years ago, and made by Australopithicine hominids, were the whole of technology for over 1.5 million years! (Acheulean tools have since been found in Africa dating back to a million years ago.) As man can be defined as the tool making animal, the development of this technology

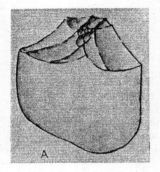

Fig. 2. Olduwan chopper or scraper. These oldest known tools date back to 2.6 million years ago. [Ref. 3, reproduced, with permission, from the (Encyclopaedia Brittanica 15th Edition Encyclopaedia Brittanica Inc., Chicago) Macropaedia Vol. 8, p. 608.]

differentiated our distant ancestors from the animal kingdom and started the process that produced Homo Sapiens, that is, us.

The Paleolithic age (Olduwan and Acheulean) was so long (about two and a half million years) that, by comparison, all later ages form only a short period, totalling ten thousand years, that could be labelled "recent". This consists of the Neolithic (New Stone Age) characterised by a variety of polished stone implements, that lasted a mere 5,000 years (the progress of materials techology was beginning to accelerate!), the bronze age of a thousand years and the Iron age running, according to Archaeologists, the 4,000 years down to the present. Obviously later subdivisions characterised by other, arguably then-dominant materials such as steel etc. could be made. This broad stone-bronze-iron succession emphasises the "serious materials" used for life-and-death survival applications as weapons and tools. Tools, in turn, are used for the construction of shelter and later for producing transport, production and power machinery. Materials for these structural and mechanical applications are required to be strong, tough and durable. Developments to satisfy these requirements are the objective of the materials discipline in most people's minds.

1.2. *The other half of the materials story*

This, however, was never the whole story. Having seen the earliest type of artifact, the Olduwan chopper/scraper, let us look next at probably the world's

Fig. 3. Gold mask on the lid of the third, innermost coffin of the boy god-king, Pharaoh Tutankhamun. (Photo copyright: The Robert Harding Picture Library, London. The actual artefact belongs to the Egyptian Museum, Cairo.)

most famous and impressive single ancient artifact (Fig. 3). This is the cover of the third, innermost coffin of the boy god-king Pharaoh Tutankhamun. Those of you who, like my family and I, saw this life size funeral portrait of Tutankhamun, when it was on display at the British Museum some years ago, will know how stunning it is, being made of solid gold and inlaid with semi-precious stones. It beautifully illustrates the fact that some materials have always been valued not for strength or practical utility, but for their appearance: their colour, transparency, and sparkle — and used for decoration, display and religious or magical purposes from ancient times. Tutankhamun's tomb dates back to about 3,300 years ago, that is, near the beginning of the iron age. Gold, which occurs in nature in the metallic state, was discovered about 5,000 years ago, probably in Mesopotamia. Silver was discovered in the copper age which briefly preceeded the bronze age proper. (Bronze is a harder, more useful alloy of copper.) So the decorative "noble" metals, gold, silver and copper, were all known before any "serious", strong metal.

Fig. 4. Naturally occurring quartz crystals. Note the imperfections in the left-hand example and the characteristic angle between the "twinned" pair of quartz crystals. The crystals on the right are of smokey (dark) quartz. This mineral can also occur in a number of other colours like rose (pink) quartz and amethyst, depending on the impurities it contains.

We also have linguistic evidence of the fascination that crystals, including gem stones, had in prehistoric times in the very word crystal. This comes from the ancient (Homeric, bronze age) Greek: krystallos which meant "clear ice". (In modern Greek krystallos has come to mean crystal, of course.) To the ancient Greeks these most characteristic and frequently occurring of large crystals (Fig. 4), were krystallos as they thought them to be ice permanently solidified by very low temperatures. This material grows, when it has space to do so, in the form shown in Fig. 4 as hexagonal cylinders topped by a distinctive pyramidal shape. It is this shape or "morphology" that is characteristic. We know this material today as quartz — one crystallographic form of silicon dioxide, of which window glass and, in highly purified form, modern optical communication fibre is made. Moreover, most of us have on our wrists "quartz" watches in which small platelets cut from crystals of quartz are excited piezoelectrically to provide accurate timing.

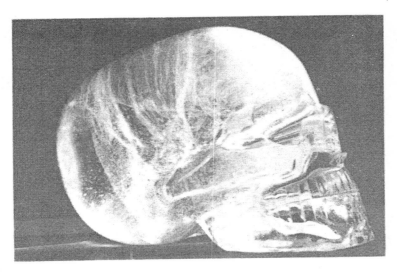

Fig. 5. The famous life size crystal skull in the Museum of Mankind in London. The filmy veils inside the material consist of innumerable small particles. (Copyright: The British Museum.)

Thus this oldest known of the physical-property, "functional" materials is at the heart of a number of the latest technological marvels.

The mystical/magical mode of thought also responded strongly to "krystallos". The so-called crystal ball, the proverbial stock-in-trade of fortune tellers is a polished sphere of quartz. Natural quartz contains defects: cracks, inclusions and filmy veils due to nonuniform impurity distributions (as in the left-hand crystal in Fig. 4 and in the large carved quartz crystal in Fig. 5). It is still believed by some that by gazing into these depths the fortune teller can see visions of the future. Both materials scientists and fortune tellers thus study defect structure in the hope of gaining the power of prediction! In the scientist's case it is the properties and performance of the sample that can be predicted. Also illustrating the religious appeal of quartz is the beautiful crystal skull (Fig. 5) displayed in the Museum of Mankind (a branch of the British Museum) in London. This was discovered in 1926 under an altar in a lost Mayan city in British Honduras and is possibly of Aztec origin.

Fig. 6. The cubic morphology (characteristic naturally occurring shape) of fluorite crystals. This is exhibited especially by the individual crystals at the top left. The insert shows fluorite crystals photographed by the characteristic blue "fluorescent" light they emit when activated by ultra-violet light. The phenomenon of fluorescence which is now used in fluorescent lamps was first discovered in fluorite and so is named for it.

Other minerals have other naturally-occurring shapes. Fluorite forms cubes (Fig. 6) and it can be excited by ultraviolet radiation to emit "fluorescence" — visible blue light. Fluorescence was named after this mineral in which the phenomenon was first observed (Fig. 6). Galena, which is mineralogical PbS (lead sulphide), grows under favourable conditions as cubo-octahedra (Fig. 7). Incidentally this is the shape of the first Brillouin zone in the electronic structure of the modern diamond- and sphalerite-structure semiconductors. The study of the shapes of crystals led to the recognition that what was invariant was the angles between the faces. This led to the Law of Rational Intercepts, Miller indices etc. and was the start of the science of crystallography. It was also observed that many crystals readily cleave into smaller blocks of particular, characteristic shapes. For example, galena cleaves into small cubes (Fig. 8). This led to another of the basic ideas of crystallography, that of unit cells, elementary volumes which fit together to form the complete crystal.

Fig. 7. Galena crystals grown on a bed of siderite and quartz. Many crystals form in this way, in among grains of other materials. The galena crystals are the ones that look metallic.

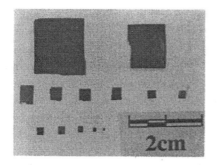

Fig. 8. Cleaved cubes of galena split from a natural crystal of this mineral.

Scientists were fascinated by the properties that crystals were found to possess. These include "fluorescence" (Fig. 6), that is the emission of cold light, now familiar through its use in fluorescent lights. Magnet is another word of Greek origin and originally referred to pieces of loadstone or magnetite (an iron oxide mineral), the first known magnetic material. Electron, too, is a Greek word meaning amber — fossilised pine resin. It was found that amber becomes electrostatically charged on rubbing. This was the first electrical phenomenon to be discovered. The first semiconductor device property, rectification, was discovered in the 19th Century by Carl Ferdinand Braun. He observed that a contact made by pressing the tip of a wire into the surface of crystals of mineral sulphides like galena (PbS) (Figs. 7 and 8), a natural semiconductor — in some places but not others! — would rectify. That is, some "point contacts" of fine wires pressed into galena pass a current in one direction but not the other. This was one of the contributions for which C.F. Braun shared the 1908 Nobel Prize with Guglielmo Marconi who pioneered radio. In the 1920s, this property was used in the first broadcast radio receivers which were known as Cat's Whisker sets, from the appearance of the fine sharpened wire contacts pressed into the natural galena crystals. Small cleavage cubes of galena can still be used very successfully as cat's whisker diodes in reception of modern radio broadcasts. Later, solid-state devices, as we would now describe galena diode radio detectors, were displaced from electronics by vacuum tubes for over 30 years. Similar silicon rectifiers were, however, used in radar in the Second World War.

This interest in silicon rectifiers lead directly to the Bell Laboratories research program on germanium, an element closely related to silicon, in the immediate post-war period. So in one sense modern solid state electronics and optoelectronics has "stone age" roots. In another sense the solid-state age began on 23 December, 1947, when the Bell Laboratories germanium research program culminated in the observation of the long-sought-after solid-state amplification, that is transistor action, for the first time. This took place in the crude point contact or "type A" transistor of Fig. 9, clearly the grandchild of the galena cat's whisker rectifier.

This first crude, experimental transistor consists of a piece irregularly cleaved out of a slice cut from a germanium crystal. This can be seen soldered to a block of metal at the bottom of Fig. 9. The metal block forms one contact to the transistor. The other two are closely spaced "point contacts" made by gold foil leads running down either side of the large triangular piece of plastic, and under the edges of its bottom tip. They are pressed down onto the germanium crystal by a paper-clip spring which is soldered to the top of a block set in the top centre of the triangle of plastic. The

Fig. 9. The first experimental type A transistor made at Bell Laboratories in 1947. The germanium crystal is the dark irregular slab resting on the yellow metal sheet forming the base contact at the bottom. The triangular block of plastic has gold foil running down its sides to form the two adjacent contacts to the top of the germanium crystal. The copper wire leads to the two top "point" contacts can be seen at the corners of the plastic triangle. The third copper wire lead is attached to the bottom metal sheet. The plastic triangle is pressed down by a rectangular spring made from a paper clip. (Photo courtesy of Bell Laboratories, Murray Hill, New Jersey.)

copper wires that lead via the gold foils to the two "point" contacts can be seen at the top corners of the triangle of plastic. The first observation of electrical amplification in this semiconductor (germanium) crystal brought Bardeen, Brattain and Shockley the Nobel Prize for Physics and launched modern solid-state electronics with all its implications.

From such physical-property, semiconductor materials are produced ever smaller "devices" that have no moving parts and do no mechanical work. It will be remembered that in elementary physics work done is defined

Fig. 10. A portion of the mill of the Analytic Engine with printing mechanism (1871) which was under construction when Babbage died. [Ref. 11, reprinted from D. Swade (1991) *Charles Babbage and his Calculating Engines* (Science Museum: London). By permission of The Science Museum.]

as the force applied times the displacement produced. No material object is displaced by electronic or optoelectronic devices, machines or systems. Instead they manipulate or "process" electrons or photons representing information. Motionless machines, like computers, television sets and air traffic control and telecomunications systems that perform no mechanical work are characteristic of the wired-society, information-technology age. This would have astounded most Victorian scientists and engineers but not Babbage! He will be remembered as the premature, lone visionary, who tried to build a programmable mechanical computer, the "analytical engine". A portion of the analytical engine of 1871, which was under construction at the time of Babbage's death is shown in Fig. 10. He was not able to complete any of his mechanical computers despite spending years of effort and advancing precision mechanical engineering along the way. Today he is recognised as a pioneer of computing and his assistant, Lady Ada Lovelace, daughter of Lord Byron, is recognised as a pioneer of the ideas of programming. The modern computer language, ADA, is named in her honour. Let me turn now to my bit of ancient history.

2. Grain Boundary Electroluminescence in GaP

In my immediate post-doctoral year at Birmingham University, I began to look around for my own field of research. I was introduced to the then new field of semiconductors by a fellow student, Dr. R.L. Bell. He had then just taken a job at the leading British laboratory for this work, the Royal Radar Establishment (now the DRE, Malvern). Professor H. Welker in Germany had just proposed on physico-chemical grounds that there should be a new class of "intermetallic" III–V compounds (Fig. 11). These, he argued, should be semiconductors, closely similar to germanium and silicon. He and his research group immediately launched a programme to synthesise and study all the III–Vs, one after another.

What lead to my entry into the field was the publication of an observation on the third compound they produced, that was, at the time, startling, at least to me. Let me fill in a little background to explain the impact of that long-ago publication. It is the common experience that clean metals are

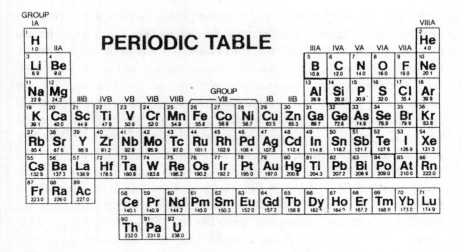

Fig. 11. The Periodic Table of the chemical elements showing group (column) IVA in which the classical semiconductors silicon (Si) and germanium (Ge) occur. (Ref. 2, reprinted from *Adv. Electron. El. Phys.*, The Physics of Semiconductor Materials, E. Burstein and P.H. Egli. **7**, 1, (1955), by permission of the the publisher, Academic Press Ltd., London.) The III–V compounds are formed of 50 atomic % each of an element from group IIIA and one from group VA, like GaAs. H. Welker first suggested that such compounds would be semiconductors closely analogous to Si and Ge and then led a research group that synthesised these materials one after the other and determined their semiconducting properties.

opaque and shiny like gold (Fig. 3) and stainless steel, for example. Insulators like silicon dioxide whether in the form of quartz (pure) or window glass are transparent or translucent. In fact one distinguishes metals from insulators or ceramics like alumina or zirconia, by eye, in this way. The dominant semiconductor material, in which transistor action had been discovered, was then germanium and the other semiconductor element, then of largely historical interest, was silicon. Both look typically silvery and metallic (Fig. 12). The naturally occurring semiconductor galena also looks darkly shiny and metallic (Figs. 8 and 9). The third of the III–V compounds that Welker's group made was gallium phosphide (GaP). They found that it was, surprisingly, transparent (Fig. 12). Moreover, when they passed a current through it, gave out light, called electroluminescence.

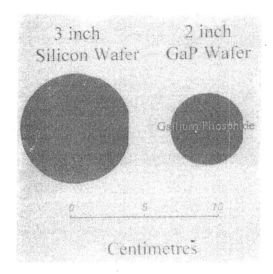

Fig. 12. Slices of shiny, metallic-looking silicon and of transparent orange gallium phosphide.

I immediately wrote to Welker saying: please, may I have some? Back came a few bits by return of post. I went and talked to Graham Alfrey, a lecturer in Physics and his research student Colin Wiggins who were interested in electroluminescence, a long-standing research field in Birmingham. We attached a couple of wires to one of the bits of GaP with dabs of silver paint, connected it across an ac voltage and saw the sort of pattern of bright lines shown in Fig. 13. We quickly did enough to verify the observation that the light came from the grain boundaries, to show that two flashes of light came out during each cycle of ac current and that the intensity increased with the magnitude of the current. We sent these fact plus the hypothesis that it was an example of injection electroluminescence in a Letter to the Editor of Nature.[7] The last point was a controversial idea at the time but it was right as it turned out. This was my first publication and I was hooked for life!

It is now clear that semiconducting materials lie toward the middle of the spectrum of optical as well as that of electrical properties. The term "*semi*conductor" was originally coined to classify materials with electrical conductivities that lie between those of good metallic conductors and of

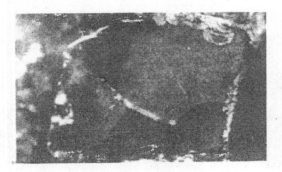

Fig. 13. Light micrograph of grain boundary electroluminescence in GaP. The bright lines show that the light emitted from this early polycrystalline GaP came from boundaries between the grains of different crystallographic orientations in the slice of material. (Ref. 1, reprinted from *Sol. State. Phys. Electron. Telecomm.* Vol. 2. Electroluminescence at Grain Boundaries in Gallium Phosphide. G.F. Alfrey and C.S. Wiggins, 747, 1960 by permission of the publisher, Academic Press Ltd., London.)

good insulators. This property depends on the arrangement of the atoms in the crystal and of the electrons in quantum mechanical states that result in what are called forbidden energy gaps — that is, ranges of energy which no electron can have, in the material as shown in Fig. 14.

In metals there are no gaps while in insulators there are wide ones. Semiconductors are materials that have intermediate gap widths [Fig. 14(a)]. Materials with gaps of the order of an electron volt or less, like germanium (Ge), silicon (Si) and gallium arsenide (GaAs) look metallic because the energies of photons of visible light ($h\nu$) are large enough ($>E_g$) so they can be absorbed to raise electrons from filled states below the gap to empty ones above it [Fig. 14(b)]. Materials with gaps above about 3 eV, however, cannot absorb photons of any wavelength in the visible range and so are transparent. GaP is on the borderline with an energy gap of 2.26 eV. It can absorb high energy (blue) photons but not medium (yellow) or low energy (red) photons so white light is partially absorbed and the nonabsorbed portion of the light is seen as yellowish or orange, depending on the thickness of the material.

The synthesis of GaP and our observation of grain boundary electro-luminescence took place in 1956–1957. Today, 35 years later, GaP is still

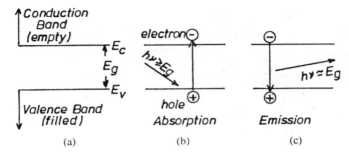

Fig. 14. (a) Electrons can only occur in energy states that lie in the ranges in the valence band (which are occupied) or in the unoccupied conduction band range of energies. These bands are separated by an energy gap of width E_g. (b) If a photon is of energy greater than the band gap (from the top of the valence band, E_v, to the bottom of the conduction band, E_c) the energy may be absorbed by raising an electron from the valence into the conduction band. This is light absorption. The resultant increase in electrical conduction can be used to detect light and to receive information encoded into the light beam. This is expoited in solar cells which generate electrical power from sunlight, photodetectors e.g. for burglar alarms or counting items moving on a production line belt and the photodetectors used in fibre optic telecommunications which receive laser beams transmitting vast amounts of information. (c) Conversely, if an electron drops in energy from a level in the conduction band to one in the valence band, the energy given up may be emitted as a photon of light of energy greater than or equal to E_g. This is the basis of electroluminescence which is the conversion of electrical current into light. It is exploited in light emitting diodes (LEDs) and semiconductor lasers.

the only "good" transparent semiconductor. That is, it is the only one of very many materials with wide enough energy band gaps to be transparent in the visible light region and which is also a well-behaved semiconductor. Its band gap, E_g, is also large enough so that, as in Fig. 14(c), it can emit photons that are of sufficiently high energy to be visible to humans. It is still the only good quality, large gap, transparent semiconductor in which we can reproducibly make *p-n* junctions that work well. The reasons for this widegap problem, namely that all other wide gap materials are difficult to make change carrier sign (i.e. to change from *n* to *p* type) by adding the appropriate amount of an impurity, is still not known. So GaP is still a rather magical material. The search for a material of even larger gap, E_g, in which "good",

 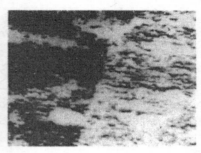

Fig. 15. Two of the earliest scanning electron microscope (SEM) voltage contrast pictures. As the voltage was reversed between the images, the active *p-n* junction (the dark/bright boundary) was displaced slightly from the one side of the the grain boundary to the other. (Look for example at the edge of the white oval to the left of the bottom of the dark/bright boundaries.) This displacement is due to the presence of an *n*-type layer between the two *p*-type grains. (Ref. 1, reprinted from *Sol. State. Phys. Electron. Telecomm.* Vol. 2. Electroluminescence at Grain Boundaries in Gallium Phosphide. G.F. Alfrey and C.S. Wiggins, 747, 1960 by permission of the publisher, Academic Press Ltd., London.)

that is cheap and reliable blue-light-emitting devices can be made (GaP absorbs the blue, remember!) still goes on. Recently it has appeared that it will probably be a sister III–V material, gallium nitride (GaN). This is not available as crystalline material of good structural quality, however. Moreover, other II–VI compounds, such as zinc selenide (ZnSe), are also competing for this ecological niche in optoelectronics.

The correctness of the mechanism we originally suggested for the electroluminescent emission from grain boundaries in GaP was clearly established by Alfrey and Wiggins after I had returned to Johannesburg and taken up a lectureship in my original Physics Department. This was done by one of the first applications of voltage contrast in a scanning electron microscope (Fig. 15). This was in 1958–1959, voltage contrast having been observed by a research student for the first time only in 1957. That occurred in the Engineering Laboratories at Cambridge where the SEM was being developed in Professor Oatley's group and where the micrographs in Fig. 15 were taken. The contrast in this observation occurred only because there <u>was</u> an *n*-type layer, almost certainly due to impurity segregation at the grain boundary while the grains on either side were *p*-type. As the polarity of the voltage reversed, the dark/bright reverse-biased *p*-*n* junction, where there is a sudden change in voltage, moved from one side of the grain boundary to the other. Hence the bright/dark (voltage jump) boundary is slightly displaced, by the width of the *n*-type layer, between the two micrographs in Fig. 15. This observation confirmed the hypothesis we had earlier proposed. Namely, that minority carriers, injected across whichever junction was forward biased, produced electroluminescence by radiative recombination with the majority carriers. Thus my first published research gave me an interest in three topics that turned out to be major themes of my subsequent research: optoelectronics, SEM methods and the "effects of defects" in devices. I was fortunate in that these areas proved to be of major and expanding interest.

I returned to Britain and joined the staff at Imperial College in 1962. As interest in scanning electron microscopy built up we organised in the later 1960s, a series of post-graduate Summer Schools on the subject. At this time the major advantage of the SEM was just beginning to be recognised. This is the fact that the electron beam dissipates its energy into a number of other forms as marked in Fig. 16(a). Each of these can be detected, amplified and

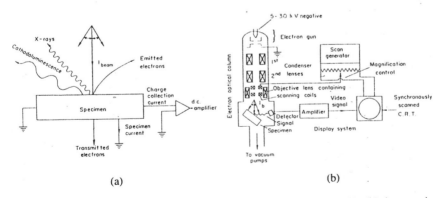

(a)   (b)

Fig. 16. (a) The forms of energy dissipation that give rise to signals used in (b) the scanning electron microscope. In (a) the charge collection current is the basis for EBIC microscopy and provides information about the electronic structure of the specimen while the cathodoluminescence is the basis for CL microscopy and provides information on the optical properties. In (b) a suitable detector receives the signal and passes it, as a current, to an amplifier before it is displayed on a cathode ray tube (CRT) to form an image i.e. micrograph.

displayed as indicated in Fig. 16(b), to act as the signal for a different "mode of scanning microscopy". The power of the SEM lies in the fact that there are these several modes and that the instrument turns each type of signal into an electrical form. This is initially displayed on a raster-scanned cathode ray tube to form a micrograph (enlarged image) in television fashion. These electrical signals are ideally suited, in addition, to data processing including image processing. It was thus an instrument whose time, technologically speaking, was most emphatically coming, as these technologies then took off.

These schools gave accounts of the quantification of the X-ray mode. This occurred through the development of convenient instrumentation and computerised correction programs. These enable one to deduce the percentage composition of the material. This was eventually refined to the point where such equipment now requires only a single button to be pressed to apply the necessary correction programs and produce quantitative readouts. This suggested to me that one should similarly work to quantify the other modes of scanning electron microscopy.

The power of the conductive and cathodoluminescence modes for the study of optoelectronic materials like GaP is clear. The conductive mode (now often referred to as EBIC — for electron beam induced current) reveals the electrical structure of the device. The principle is that energy is absorbed from the beam to raise electrons from energy states in the filled valence band to ones in the empty conduction band, again as shown in Fig. 14(b), creating an "electron–hole pair" of charge carriers. Thus the incident electron beam creates hundreds or thousands of electron–hole pairs for each high-energy incident electron. Where there is a built-in electric field in the crystal, these carriers, having opposite charges, will be sent in opposite directions (Fig. 17). This creates a current pulse that can be detected. This happens each time the scanning beam falls on or near an electrical barrier and on a viewing screen such features appear bright. Hence in semiconductor devices the "internal working parts", the *p-n* junctions, Schottky barriers and heterojunctions, are made visible. This mode of microscopy is now widely used for inspection. Electrical measurements can also be made with good spatial resolution and any electrically active defects can be seen.

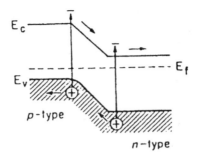

Fig. 17. Energy band diagram of a *p-n* junction under electron bombardment creating electron–hole pairs as shown by the vertical arrows. The electrons (−) are accelerated to the right by the "built-in" field at the junction while the holes (+) are sent to the left. This sending of the different charges to opposite sides of the junction is called "charge collection". It results in a so-called electron beam induced current (EBIC) when the beam of a scanning electron microscope is incident on or near a *p-n* junction or other electrical barrier in a semiconductor.

The cathodoluminescence (CL for brevity) that is, cathode-ray-bombardment-induced light, reveals the luminescence properties and structure of the material or device. The word cathodoluminescence is unknown to the public but it is in fact the most widely observed of all technological phenomena. The emission of light by cathodoluminescent coatings when a scanning electron beam falls on them is the means by which TV pictures are emitted. It is also the means by which the information is displayed on computer monitor screens, as well as in the images seen in electron microscopes of all kinds. Thus CL is of great importance in its own right. In 1968, Cambridge Instruments were finally persuaded to put the SEM into commercial production for the first time and I applied for a grant and obtained an early machine, numbers 13 or 14 in the world. SERC was already learning its modern tricks and only gave us about 2/3 of the price of this "Stereoscan" machine. We were fortunate in finding a cooperative group in the Geology Department who were happy to put up the rest of the money and join in the work. So, from the beginning, part of the time of our machines has gone to the study of "stone age" materials and this interest is continuing today, in a sense, with research programs on ceramics including the exciting new high temperature superconducting ceramics.

3. Scanning Electron Microscope Electron Beam Induced Current (EBIC) and Cathodoluminescence (CL)

Ever since the arrival of our first machine, near the end of 1968, we have worked on the development of these EBIC and cathodoluminescence modes of the SEM as quantitative analytical i.e. measurement techniques. Happily, our work on detection systems has been taken up commercially. Our EBIC detection system was evolved through four successively improved designs by people in my group guided by my former colleague Dr. Bhiku Unvala, whose company later developed it as the commercial Matelect Induced Signal Monitor ISM-5. The instrument also allows operation via software from the keyboard of a microcomputer image processor, so quantitative line scans can be recorded directly. We have found it invaluable in our work for a great variety of purposes and I understand that many these instruments have been sold, mostly for export (Fig. 18). Figure 19 shows a typical high resolution

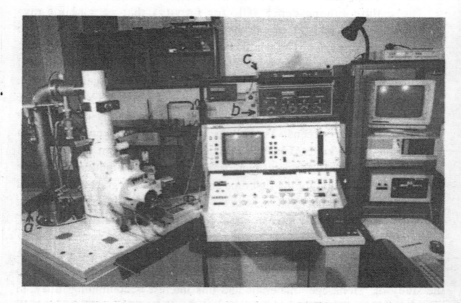

Fig. 18. The SEM and EBIC system. (a) is the head amplifier, (b) the main control unit and (c) is the computer interface module of the EBIC system.

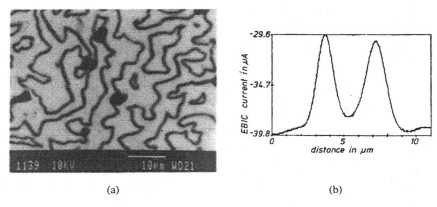

(a)   (b)

Fig. 19. (a) EBIC image of antiphase boundary (APB) structures in a GaAs/Ge solar cell for use in space satellites. (Ref. 9, reprinted from *Mat. Sci. Eng.* **B42**, D.B. Holt, C. Hardingham L. Lazzarini, L. Nasi, C. Zanotti-Fregonara, G. Salviati and M. Mazzer, Properties and Structure of Antiphase Boundaries in GaAs/Ge Solar Cells, 204, 1996 with permission from Elsevier Science.) (b) A quantitative record of the variation of the EBIC current along a line scanned across a pair of such APBs. (Ref. 4, reprinted from Microscopy of Semiconducting Materials 1997 Conf., Series No. 157. C. Hardingham, D.B. Holt, L. Lazzarini, M. Mazzer, L. Nasi, B. Raza and C. Zanotti-Fregonara, Antiphase Boundaries in GaAs/Ge Solar Cells, 573, 1997 with permission from the Institute of Physics, Bristol.) The apparent peaks actually represent reductions in the negative EBIC current (plotted downward) corresponding to the dark APB lines. By analysing these current minima, it is possible to determine both the interface recombination velocity of the APBs and the values of the minority carrier diffusion lengths on either side of them.

image and the profile of the signal recorded as a line is scanned, obtained using this equipment. Line scan profiles are used in quantitative analyses of EBIC information.

For CL too, we developed a series of successive instrument designs through the work of a few generations of research students and research assistants before we built our first liquid-helium CL stage system. This was designed by Phil Giles, based on an Oxford Instruments continuous flow cryostat that can cool the specimen to liquid helium temperatures or run on liquid nitrogen for less demanding problems. It worked so well that I suggested to the Oxford Instruments sales engineers that commercial exploitation of the CL stage might gain them a new market. Fortunately, as it happened, Oxford Instruments were then looking at the electron microscopy market. In those long past times we naively drew up a simple royalties agreement with Oxford Instruments. This netted us some money over the next four or five years, part of which we took in the form of their commercial CL stage. This is shown in Fig. 20. The cryostat and CL light collection system of mirrors are the black unit attached to the side port of the SEM. Figure 21 shows a typical CL image of misfit dislocations in a heterojunction in one of the quantum well solar cells which were first proposed by and are now under study by Dr. K. Barnham's group in the Physics Department at Imperial College.

Oxford Instruments have continued development and eventually put a complete CL cryostat stage and computerised monochromator system on the

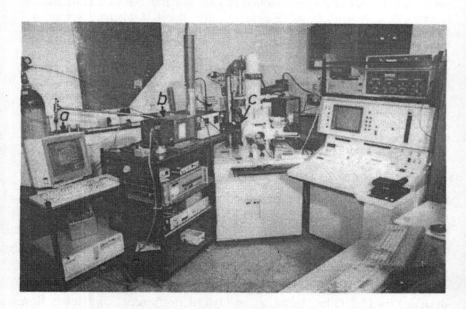

Fig. 20. The CL system. The trolleys at the left hold (a) the control and data logging computer, (b) the CL monochromator system that we developed and, mounted on the side port of the scanning electron microscope, (c) is the Oxford Instruments liquid helium cryostat and CL collection stage unit.

Fig. 21. CL image of a misfit dislocation network in an interface of a quantum well GaAs/ GaAlAs solar cell. Again quantitative line scans can be analysed to derive the recombination strength of the dislocations. (Ref. 10, reprinted from *Mat. Sci. Eng.* **B42**, M. Mazzer, E. Grunbaum, K.W.J. Barnham, J. Barnes, P.R. Griffin, D.B. Holt, J.L. Hutchison, A.G. Norman, J.P.R. David, J.S. Roberts and R. Grey. Study of Misfit Dislocations by EBIC, CL and HRTEM in GaAs/InGaAs Lattice-strained Multi-Quantum Well *p-i-n* Solar Cells. 43, 1996, with permission from Elsevier Science.)

market called the *monoCL* (for monochromator cathodoluminescence) system. It is based on the same Bentham Instruments monochromator that we had settled on, as giving a good balance of resolution and throughput, in a lightweight box but now mounted permanently on the scanning electron microscope (SEM) port. The Semiconductor Interdisciplinary Research Centre at Imperial College has acquired a monoCL so the use of SEM-CL for materials and device studies is expanding here. It turned out that there was a significant market for CL systems for SEMs and, over the years, they have sold well, mainly overseas.

3.1. *Monte Carlo electron trajectory simulations*

To quantify EBIC and CL we use Monte Carlo electron trajectory simulation programs running on microcomputers. Monte Carlo methods were introduced by von Neumann for solving problems arising in the war-time atomic bomb program. In the late 1950s and early 1960s Monte Carlo programs run on mainframe computers by people working in Cambridge helped solve problems involved in quantifying X-ray microanalysis, i.e. the X-ray mode of the SEM, then an active research topic. An American on sabattical leave in Cambridge, became enthused with the idea and on his return to the US National Bureau of Standards, he and his colleagues wrote a simplified Monte Carlo program in 1976 that could be run on the microcomputers that were then first becoming available.

This was taken up by David Joy in 1982 for application to the interpretation of SEM data. I spent six weeks working with Dr. Joy at the Bell Telephone Laboratories in Murray Hill, New Jersey in 1986 and myself became a convert. He kindly gave me a copy of his programs on floppy disc. On my return, I in turn got my Research Assistant, Dr. Eli Napchan interested. So this work has oscillated across the Atlantic repeatedly!

Dr. Napchan has written a suite of menu-driven programs evolved from the Cambridge — N.B.S. — Joy programs that he now calls MC-SET (Monte Carlo Simulation of Electron Trajectories). MC-SET is adapted to deal with epitaxial multilayers of III–V materials that are characteristic of optoelectronics, in laterally limited device structures. It now includes menu-driven options to calculate EBIC (electron beam induced current i.e. the conductive mode of the SEM) and CL signal strengths and image contrasts. We know the programs are useful because they are made available freely and Dr. Napchan has had well over 100 written requests from people to whom he has sent the program on disc. Outside collaborations have also led to papers coauthored by him.

The way in which Monte Carlo programs work is illustrated in Fig. 22. The basic assumption is that the interactions of the fast beam electrons are either elastic or inelastic. The inelastic processes cause the beam electron to steadily loose energy without changing direction significantly whereas elastic collisions with the nuclei of the atoms scatter the electron through relatively large angles without losing energy. For each elastic collision, the computer generates two random numbers. One is used to calculate the scattering angle ϕ and the other to determine the azimuthal angle ψ. This use of random numbers is the reason for the gambling-related name, "Monte Carlo" for such methods. The type of display of the result that is produced in a very

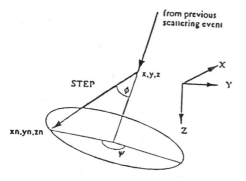

Fig. 22. Monte Carlo simulated trajectory steps and deflection and azimuthal angles. (Figs. 22 to 25 are reprinted from Scanning **16**. D.B. Holt and E. Napchan, Quantitation of SEM EBIC and CL Signals Using Monte Carlo Electron Trajectory Simulations, 78, 1994 (Ref. 6). Copyrighted and reprinted with the permission of SCANNING and/or the Foundation for Advances of Medicine and Science (FAMS), Box 832, Mahwah, New Jersey 07430, USA.)

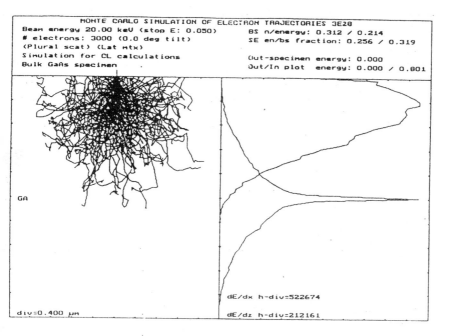

Fig. 23. MC-SET produced display showing, on the left, the calculated trajectories (paths in the solid) for the first 200 electrons of a beam of 25 keV, incident normally on a GaAs specimen. On the right are plotted the "depth dose" i.e. the energy deposited per unit depth (the upper curve) and the "lateral dose" i.e. the energy deposited in unit radial distance from the beam impact point, at the peak of the lower curve.

short time by running this program for a particular type of specimen and set of SEM operating conditions is shown in Fig. 23.

The results of such a calculation can be used, for example, to find the intensity of the CL emitted from simple bulk GaAs as a function of beam energy. For the calculations the current is kept constant so the number of electron trajectories simulated for each beam energy is constant. Hence, also, the power input increases linearly with the beam energy in keV and, at first, so does the intensity of the CL. However, the beam also penetrates deeper, so the photons have increasing distances to travel to reach the top surface and escape. So self-absorption in the material reduces the CL intensity giving the form of curve shown in Fig. 24. The effect of varying the assumed value of the absorption coefficient, α, is also shown and by fitting experimental data to the calculated curves, a value for α can be obtained.

In addition there are found to be effects that are ascribed to surface "dead layers". These layers are nonluminescent because built-in fields, due to surface charging, separate the hole–electron pairs generated by the beam and prevent them recombining to produce light as in Fig. 14(c). The effect

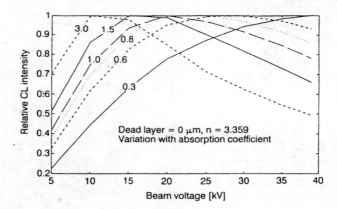

Fig. 24. MC-SET calculated curves of the variation of the emitted CL intensity for the values of the incident beam energy plotted along the *x*-axis and a constant beam current for a solid specimen of GaAs. Each curve represents the intensity variation to be expected for the particular value of the self absorption coefficient marked for it. (After Holt and Napchan, 1994.)

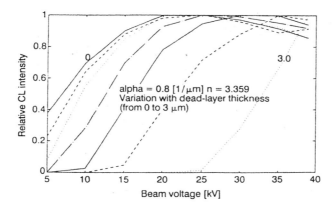

Fig. 25. The effect of increasing dead layer depths, *d*, as marked, for curves like those of Fig. 24. (After Holt and Napchan, 1994.)

of different values of dead layer depth, *d*, is shown in the set of curves in Fig. 25. When this dead-layer depth is greater than the effective beam penetration range no light is observed but, after that, the curve takes a similar form to that simulated before. Again curve fitting experimental results can determine this depth. Such quantitative interpretation of CL measurements was not previously possible and such data can now be calculated in a few minutes. Hence we can understand the enthusiasm with which the Monte Carlo method has been welcomed in the SEM world.

3.2. *Applications of scanning electron microscopy to optoelectronic devices*

Applying these techniques, over the years, to problems in microelectronic and optoelectronic devices has given us a grandstand view of development from the first practically useless optoelectronic devices to the awesome achievements of the present.

When we got our first SEM we ·had a research link with the then Admiralty electronics research laboratory SRDE (Signals Research and Development Establishment) supporting a student, Brian Chase, to look at materials problems in the prototype GaP light emitting diodes (LED's) of the time, including means for recognizing *p-n* junction material that would make efficient LEDs without having to process and test the diodes (an expensive business).

For a number of years III–V research had been kept going only by the fact that workers at the Philips Laboratory in Eindhoven had once, I believe literally once only, made a red LED that was 1% efficient. It took years to learn to make even more efficient ones reproducibly. However, the GaP LED was, historically, the device that eventually took III–V technology into production. These devices arrived on the market about ten years after our original observation of grain boundary EL, just in time to satisfy the first economic demand for compact electronic displays as shown in Fig. 26. This too is an historic artefact dating back to the days when Britain had a domestic consumer electronics industry. It is the Sinclair Cambridge hand calculator that my daughter used for years. Hand calculators and electronic watches both needed displays and the GaP (and GaAsP) LEDs provided this for the first several years. This provided a market for millions of devices per annum and first put the III–V optoelectronic materials on the industrial map! In the middle of Chase's PhD work, SRDE suddenly said they were no longer

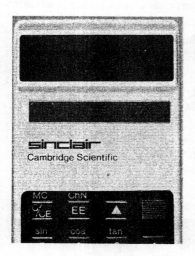

Fig. 26. A Sinclair Cambridge hand calculator showing the red GaP LED α-numeric display.

interested in GaP LEDs because these devices were leaving the research stage for initial commercial production. So, they said, will you please work on GaAs lasers. These had then just been produced following theoretical indications of the practicality of the device — by three US laboratories and, independently, but a couple of weeks later, by SRDE.

The first GaAs lasers were very large, simple devices being just *p-n* junctions, very heavily doped on both sides. They were of the thickness of a wafer, about 0.4 mm, and around 1 mm by 2 mm in area (Fig. 27). When forward biased to pass enormous current densities (of the order of 10^5 A/cm^2), the diodes emitted the fascinating, new, coherent but invisible infrared laser light. Cynics described them as "solutions in search of a problem" since there was no known or obvious application for such a thing.

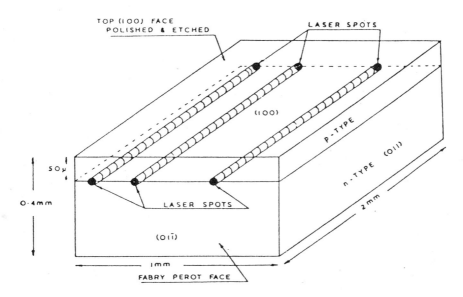

Fig. 27. Schematic diagram of a GaAs *p-n* junction laser showing the filaments that were found to begin laser emission, one after the other, as the current through these devices was increased above the threshold value. (Ref. 5, reprinted from *J. Mat. Sci.* **3**. M.J. Hill and D.B. Holt, Defects and Laser Action in GaAs Diodes. 244, 1968. Published by Chapman and Hall.)

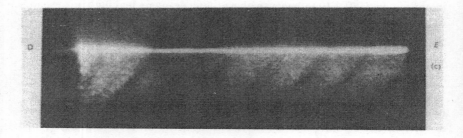

Fig. 28. Transmission infrared light micrograph of the emission face of one of the first GaAs *p-n* junction lasers. Also shown is the emission along the junction line D to E from a number of large bright spots mainly near D. These are the filamentary laser emission sites, characteristic of these first crude devices. It can be seen that the filamentary emission spots occur between the inclined dark infrared absorption bands that can be seen below the junction. These impurity striations arose during the growth of these GaAs crystals. (Reprinted from same source as Fig. 27.)

There were severe drawbacks to these early devices. They could only be operated intermittently when cooled to liquid nitrogen temperature, on for microseconds and off for milliseconds. That is they would emit only for about a thousandth of the apparent lasing time and then they only lasted for tens of hours. The very large currents burned them out in a few hours so they had real operating lives of seconds of total emitting time. The other problem with these devices was multimode emission. As the current was turned up through the threshold value, first one filament, as shown in Fig. 27, and then another began to lase, independently i.e. not coherently with each other. Even single filaments emitted complex sets of beams in slightly different directions. Martin Hill carried out a PhD on the problem and showed it to be due to the presence of impurity growth striations in the "bulk" GaAs material of the wafers as shown in Fig. 28.

These problems were cured by the design of "double heterostructure (DH), stripe geometry" lasers (see Fig. 29). These incorporated two heterojunctions i.e. interfaces between three layers of materials with different bandgap widths [see Fig. 14(a)]. These were AlGaAs, GaAs and AlGaAs in the first DH lasers as shown in Fig. 29(a). These materials were so chosen that the resultant steps in the band edges [see Fig. 29(b)] confined the

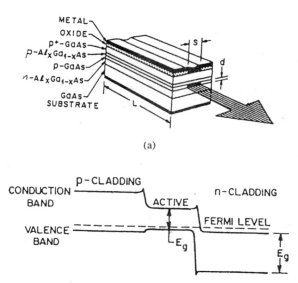

(a)

(b)

Fig. 29. (a) DH (double heterostructure) laser layer structure showing the top contact stripe of width *s* and the active layer of depth *d* and (b) its energy band diagram.

injected carriers in the active layer where they could produce (laser) light and also confined the photons in this layer to remain part of the laser beam. The use of a conducting stripe [the strip of width S on the top of the device in Fig. 29(a)] also meant that current only passed down through a narrow vertical slab. This further reduced the total current required and so increased the working life of these devices. It also ensured that only one lasing filament could occur. The final result was that the lasers were better and lasted much longer.

The DH laser design was just one of a multitude of device possibilities opened up by the development of epitaxial growth technology. Epitaxy is the growth of one crystal on another in some simple, well-defined orientation relation. When growth takes place on a single crystal substrate, the deposited film is also single crystal material. Even today the best bulk III–V crystals are of inadequate quality to make electronic components. However, they can be cut into slices on which epitaxial layers of good "device" quality can be grown. In fact down to the present all III–V devices are produced in epitaxial layers, usually in structures of epitaxial multilayers like that of the pioneering DH laser.

These early DH lasers were able to operate, continuously on, at room temperature. To extend and guarantee operating life a good deal of research had to be done on the ways in which rapid failure occurs. This knowledge could then be used to prevent these mechanism from operating. Two modes of rapid (thousands of hours) failure were found. One of these is shown in the micrographs of Fig. 30, taken by one of our students working on samples provided by Bruce Wakefield from British Telecoms Research Laboratories. Wakefield showed that this mode of failure was due to rapid point defect migration down through the active stripe during laser operation under the influence of the elastic strain field that could arise at the edges of the stripe. When this and the dark line defect (DLD) mode of failure had been correctly diagnosed and cured, the operating life of these first relatively simple GaAs-based DH lasers reached 10^5 hours i.e. about ten

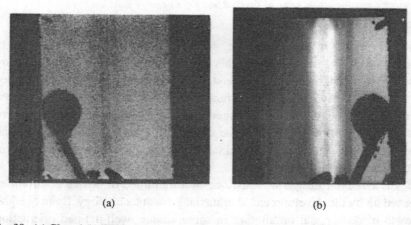

(a)   (b)

Fig. 30. (a) CL and (b) EBIC micrographs looking down through the structure of a Wakefield-failed DH laser showing the luminescence and electrical deterioration of the active central stripe of the active layer.

years which sufficed to enable them to be used in the first generation of fibre optic telecommunications lines.

A bewildering range of laser device structures have since been designed and evaluated first in GaAs- and then in InP-based alloy multilayer structures. We have long cooperated with workers in the laser group at the Caswell industrial laboratory, originally belonging to the Plessey company but now renamed GEC-Marconi Materials Technology Limited. One of the more interesting designs of laser they asked us to look at was the later InP-based Double Channel Planar Buried Heterostructure (DCPBH) laser with the form of structure shown in a 1986 British Telecom Laboratory illustration in Fig. 31. In this and all later laser designs, instead of a narrow conducting stripe there is only a narrow ridge of the laser structure left after etching away the epitaxially grown layers on either side. The sides of the ridge then

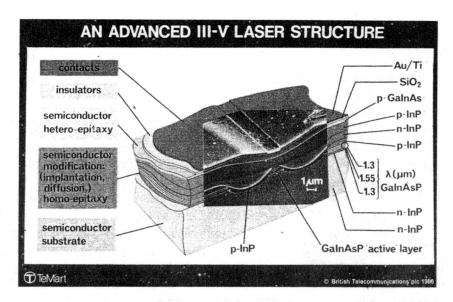

Fig. 31. The schematic diagram of a double channel planar buried heterostructure (DCPBH) laser. The "buried heterostructure" is that of the laser layers in the central region. The "double channel" refers to the *p-n* alternations in the channels filled in on either side of the central section. (Reprinted with the permission of British Telecommunications.)

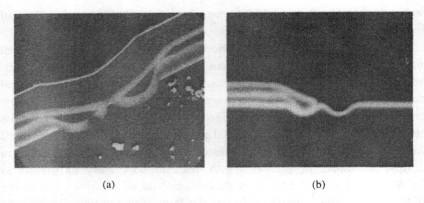

(a)   (b)

Fig. 32. SEM EBIC images of DCPBH lasers showing (a) a short circuit path (the black dot interruption) through one of the bright *p-n* junctions in one DCPBH laser and (b) a completely missing *p-n* junction in another. A perfect electrical structure would appear like a symmetric pair of half-lens spectacles.

have to be protected in some way. In this design the grooves are filled in by further epitaxial growth of material containing a couple of curved *p-n* junctions to block any leakage current bypassing the active layer stripe of the laser. They appear as bright lines in SEM EBIC images in these devices in a form like half-lens spectacles. Our examination of such lasers showed that it is easy for the epitaxy to go wrong so the *p-n* junctions are malformed as shown in Fig. 32. Plessey concluded that this interesting but complex structure was a bit too tricky to produce economically.

These few examples of observations on successive types of laser will serve as snapshots of trouble shooting studies during the decades of broadly based progress in this field. This has produced in the operating life of semiconductor lasers what is, as far as I know, the largest factor of improvement in the history of technology. We saw that the real operating ("on") life of the first GaAs injection laser diodes was of the order of seconds, while the first generation of DH stripe geomentry lasers had lives of 10 years. The next, late 1980s generation of InP-based lasers simply did not undergo either of the rapid types of deterioration found in the DH GaAs-based devices and their lives were soon pushed up to 10^6 hours which is over a century. This is already an increase in working life of a factor of about 10^9 times, that is, about an (American) billion times improvement. I know of no other artifact that can claim anything similar. For comparison take the well-known case of Si chips. The first integrated circuits were produced in 1960, five years before the first semiconductor lasers and contained a few components. Now, in the year 2000, the largest chips in production contain a few tens or hundreds of millions of components a "mere" ten or a hundred million times increase!

In 1965, lasers were a solution in search of a problem. Today, a quarter century later, the world's vastly expanded telecommunications traffic (e-mail, Internet and all) is transmitted on injection laser beams along silica fibres and many millions of semiconductor lasers per annum are bought by the public in CD record players.

4. Scanning Electron Microscope Characterisation of Quantum Confined Structures

Now let me touch on examples of a final optoelectronic phenomenon we have studied that will bring us down to present day technology. One of most striking developments of recent years is the field of low dimensional solids. Esaki and Tsu in 1970 suggested that it would be possible, using epitaxial technology, to grow alternating layers of different semiconductors of thicknesses less than the mean free path of electrons or the electron wavelength and predicted interesting new energy band structures would occur in the artificial superlattices so formed. These should result in new artificial, man-designed and man-made crystal properties.

The first attempts, made by Blakeslee in Esaki's IBM Laboratory, using vapour phase epitaxy (VPE) succeeded in making the structures but they did not exhibit the predicted new properties. Amusingly, however, IBM Laboratories offered these structures as resolution test specimens since the visible layer thicknesses could be made around the limit of spatial resolution of the first SEMs (about 15 nm). I have always regretted that I did not buy one of those samples.

The technique of molecular beam epitaxy (MBE) was then adopted and developed for this work. Esaki's group and others, including the group that became the nucleus of the Semiconductor Interdisciplinary Research Centre at Imperial College, then suceeded in growing artificial crystals with new "designer" energy band structures that indeed exhibited predictable new effects. This created a vast and most attractive new field of materials research and practical applications in micro- and opto-electronics.

One of its most attractive features is the simplicity of the Physics involved. The simplest "quantum confined" or low dimensional structure is the single quantum well (SQW) of three epitaxial layers of two different

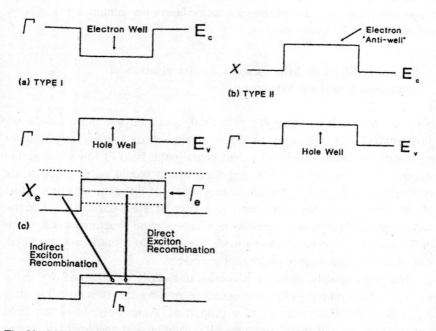

Fig. 33. SQW (single quantum well) energy band diagrams. (a) A type I single quantum well, (b) a type II quantum well and (c) the possible forms of direct and of indirect exciton recombination that can both give rise to luminescent emission from a type II well. (Figs. 33 to 35 are reprinted from *Mat. Sci. Eng.* **B9**. D.B. Holt, C.E. Norman, G. Salviati, S. Franchi and A. Bosacchi. Type II Indirect and type I Direct Recombinations in GaAs/AlAs Single Quantum Wells. 285, 1991 (Ref. 8).

energy band gaps as shown in Fig. 33(a). This can be modelled, to a first approximation, by the simplest problem in quantum mechanics, known as the "particle in a one-dimensional box". This type of simplicity led Esaki to describe the field as do-it-yourself quantum mechanics. Such simple quantum mechanics tells us that there are trapped electron and hole states in both wells as shown by the horizontal lines in Fig. 33(c). They are connected by arrows that indicate possible light-emitting electron quantum jumps. The energy levels of these trapped states depend strongly on the width of the well i.e. on the thickness of the middle epitaxial layer.

This has two important consequences. (1) The energy levels are separated by a distance (energy) that is different from that of the energy band edges of the bulk material so a photon of different energy is emitted. That is, the wavelength of the light can be made shorter (or longer in the case of a type II quantum well). Hence the wavelength emitted can be tailored for particular applications simply by growing a well of the required width and type. (2) The electrons injected from one side and the holes from the other become trapped in the same thin layer so they interact strongly. Consequently, the luminescent efficiency of QWs is high. We knew that in principle but, when we looked at our first SQW in the SEM, we were still astonished. That was in material grown by a group with whom we work closely at the MASPEC Institute in Parma, Italy. The structure is shown in Fig. 34. The important bit is the single 3 nm (about 10 atoms) thick layer near the top with the interfaces marked A and B. When the electron beam fell on this we found, on looking

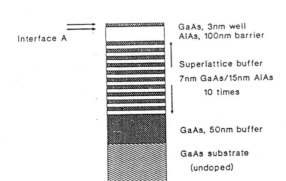

Fig. 34. Schematic diagram of the MASPEC quantum well specimens. (After Holt *et al.*, 1991.)

into the turret where the CL comes out, that a bright red light was visible like that from a red GaP LED such as we saw earlier. However the beam excites only an area about a micron by a micron whereas the LED is of the order of a mm by a mm i.e. a million times larger. Moreover the SQW is 3 nm thick whereas the LED emits from a thickness of about a micron i.e. nearly a thousand times greater. Thus the emission density i.e. the source volume brightness of the SQW was anything up to 10^9 times greater than that of an LED! Moreover, the GaAs of which the SQW layer is made cannot, in its natural bulk state, emit light visible to human eyes at all. To use a favourite phrase of my old mentor, Professor Nabarro, the effect is not great, its enormous! Its enough to make one believe in quantum mechanics (something some of our students find difficult)!

Obviously quantum wells are ideal for use in the active layer of injection lasers because of their high efficiency and brightness on the one hand and their designer tailored wavelengths i.e. emission colours on the other. Quantum wells are in fact incorporated in a high proportion of the laser designs of recent years.

The MASPEC samples had another interesting feature. The energy band diagram of the SQW was not of the simplest type I in Fig. 33(a), but of the form known as type II [Figs. 33(b) and 33(c)]. What we found was that at low beam energies, 1 keV, the CL emission spectrum from the SQW (the top 3 nm well in Fig. 34) had the form shown on the left in Fig. 35(a) which occurs when only the lower indirect recombination states are filled and give rise to light emission. At the higher energy of 3 keV, the light emitted by the SQW changed to the much broader and shifted form shown at the left in Fig. 35(b). This type of SQW emission we attributed to the beam of higher energy filling all the lower-energy indirect states so the higher-lying, direct states also begin to fill. The direct exciton recombination produces additional light at a higher energy and shorter wavelength as in Fig. 33(c). The addition of the direct emission at higher energy to the indirect at lower photon energy produces the shift and shape change in the 3 nm SQW CL emission band. In contrast, the emission from the deeper MQW and bulk GaAs does not change on going from a beam energy of 1 to 3 keV (except to become stronger).

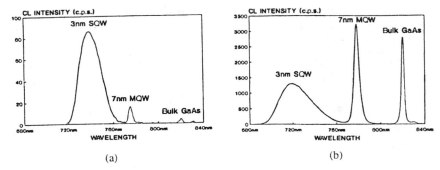

Fig. 35. CL emission spectra from a specimen of the type shown in Fig. 34 at beam energies of (a) 1 keV and (b) 3 keV. SQW marks the light from the top 3 nm single quantum well of interest, the MQW (multi QW) luminescence is that from the 7 nm wells of the "superlattice buffer" and the bulk GaAs emission comes from the "buffer" and "substrate" material at the bottom of Fig. 34. The SQW band is symmetric in (a) but asymmetric in (b). (After Holt *et al.*, 1991.)

Thus our survey has now taken us from the materials of 2,600,000 years ago to some of those of today.

5. Magical Materials for Motionless Machines

Now where has all this research and development landed us? From the viewpoint of a Victorian observer familiar with 19th Century industrial developments, present day trends would appear most strange as suggested by my title: magical materials for motionless machines. The first industrial revolution applied and is still applying strong materials to generate and harness ever increasing power, first steam then other forms, to drive machines ever faster, to take over productive work and replace muscular effort.

Society is now in the early stages of a second industrial revolution for which the term information technology is used. The typical machines of this revolution sit motionless and do no work but manipulate electrons (electronics) or photons (this technology is called photonics or opto-electronics) which encode information.

We have mentioned as an application of our materials the semiconductor lasers that launch the signals down the fibre optic telecommunication links that form the nervous system of world society. Cathodoluminescence, although the word is unknown to the public, is the means whereby TV screens emit their pictures and computer monitors their information. Silicon chips, which I have not talked about here, but which we also work on, have given us computers and smart, automated machines, from domestic washing machines to robots. Computers linked via optical fibres give us e-mail, fax, instantaneous worldwide trading and travel booking, international credit cards and cash machines. The impact of TV on, for example, American opinion on the Vietnam War or on people's environmental awareness and concern for the victims of famine in Africa and war in Yugoslavia has been enormous.

Moreover, research is now in progress on artificial intelligence. At present, as baffled users know, computers have only artificial stupidity! If machines with artificial intelligence do come into existence, physical materials will have come to be used in devices and systems that, in some sense, think. If (or when?) that happens, these materials, once regarded as magical, will have proved marvellous indeed!

Acknowledgements

I would like here to acknowledge the work of all the members of my research group over the years although they are too numerous to mention individually.

As a tribute to our too often forgotten patrons, the British taxpayers, let me leave an appropriate last word on research to Sir Winston Churchill. He wrote of Christopher Columbus's voyage of discovery to the New World that: "He left without knowing where he was going, arrived where he did not think he would, and all of it at the expense of others." Now that's what I call research and I have enjoyed a lifetime of it!

References

1. G.F. Alfrey and C.S. Wiggins, Electroluminescence at Grain Boundaries in Gallium Phosphide *Sol. State. Phys. Electron. Telecomm.* **Vol. 2**, Academic Press, London, 747–750, 1960.

2. E. Burstein and P.H. Egli, The Physics of Semiconductor Materials. *Adv. Electron. El. Phys.* **7**, 1, 1955.

3. Encyclopaedia Brittanica 15th Edition (Encyclopaedia Brittanica Inc., Chicago).

4. C. Hardingham, D.B. Holt, L. Lazzarini, M. Mazzer, L. Nasi, B. Raza and C. Zanotti-Fregonara, Antiphase Boundaries in GaAs/Ge Solar Cells in Microscopy of Semiconducting Materials 1997 Conf., Series No. 157 (Institute of Physics, Bristol), 1997.

5. M.J. Hill and D.B. Holt, Defects and Laser Action in GaAs Diodes. *J. Mat. Sci.* **3**, Chapman and Hall, 1968, 244–258.

6. D.B. Holt and E. Napchan, Quantitation of SEM EBIC and CL Signals Using Monte Carlo Electron-Trajectory Simulations. *Scanning*, **16**, 78, 1994.

7. D.B. Holt, G.F. Alfrey and C.S. Wiggins, Grain Boundaries and Electroluminescence in Gallium Phosphide. *Nature*, **181**, 109, 1958.

8. D.B. Holt, C.E. Norman, G. Salviati, S. Franchi and A. Bosacchi, Type II Indirect and Type I Direct Recombinations in GaAs/AlAs Single Quantum Wells. *Mat. Sci. Eng.* **B9**, 285, 1991.

9. D.B. Holt, C. Hardingham, L. Lazzarini, L. Nasi, C. Zanotti-Fregonara, G. Salviati and M. Mazzer, Properties and Structure of Antiphase Boundaries in GaAs/Ge Solar Cells. *Mat. Sci. Eng.* **B42**, 204, 1996.

10. M. Mazzer, E. Grunbaum, K.W.J. Barnham, J. Barnes, P.R. Griffin, D.B. Holt, J.L. Hutchison, A.G. Norman, J.P.R. David, J.S. Roberts and R. Grey, Study of Misfit Dislocations by EBIC, CL and HRTEM in GaAs/InGaAs Lattice-Strained Multi-Quantum Well *p-i-n* Solar Cells. *Mat. Sci. Eng.* **B42**, 43, 1996.

11. D. Swade, Charles Babbage and his Calculating Engines (Science Museum: London), 1991.

Professor Alan Atkinson

Alan Atkinson joined the Department of Materials in 1995 from AEA Technology (Harwell) where he was head of Materials Chemistry Department. After graduating from Cambridge he studied the electronic properties of metals for his PhD at Leeds University. He then investigated the processing of silicon nitride ceramics before moving to Harwell in 1975. There his research interests included: mass transport in ceramics (particularly at grain boundaries); high temperature corrosion; sol-gel processing of ceramics; cements and concrete for the disposal of radioactive waste; catalysts and adsorbents for environmental pollution abatement; and the mechanical properties of thin films. His current research topics include solid oxide fuel cells, mechanical properties of thin films and cement chemistry. He is a Fellow of the Institute of Materials and the Institute of Physics and was awarded the Carl Wagner Prize for his work on high temperature corrosion. He has published over 140 papers in scientific journals.

INTERFACES IN MATERIALS — IF YOU CAN'T BEAT THEM, JOIN THEM

A. ATKINSON

Department of Materials
Prince Consort Road
London, SW7 2BP, UK
E-mail: alan.atkinson@ic.ac.uk

1. Introduction

Interfaces in materials play an important role in determining the properties of materials and structures and in enabling their fabrication. They are usually just as important as the intrinsic properties of the materials themselves and often more so. They can be grouped conveniently as: grain boundaries between crystals of different orientation in polycrystalline materials; interfaces between two different solid materials; or interfaces between solids and liquids. (The interface between a solid and a gas, or vacuum, will not be considered here, since this constitutes the separate field of surface science.)

Interfaces usually have different arrangements of atoms and different properties from bulk materials with the result that interfaces are often a source of unwanted problems. For example they can: be preferentially attacked chemically; accumulate impurities; provide fast diffusion pathways; be mechanically weak; or interfere with electrical conduction. In a few cases it has proved feasible to avoid such problems by developing materials that have no interfaces; that is, single crystals. For example, single crystals of silicon are used on an enormous scale for the manufacture of integrated circuits in order to avoid the electrical problems arising from grain boundaries. Similarly, the latest generation of turbine blades for high performance jet engines are made from single crystals of "superalloys" to avoid the mechanical weakness of grain boundaries at high temperatures. But using single crystals is an expensive strategy that is only practicable in a small number of applications. In the majority of cases the problems arising from interfaces are being solved by improving understanding and then modifying their properties to bring them under control. Furthermore, the properties of interfaces are not all bad. As we shall see, interfaces in materials are providing an ever increasing spectrum of opportunities in which their properties can be exploited beneficially.

The field of interfaces is very large and here I shall consider only a very small part of it drawn from areas in which I have been personally involved. Nevertheless, the examples given should serve to illustrate the level of understanding that has been achieved, the ways in which this is being exploited and the exciting potential for the future. I will begin by summarising the way in which our understanding of grain boundaries has developed historically and follow this with an illustration of the role played by oxide grain boundaries in the high temperature corrosion of metals. I will then discuss some aspects of the mechanical properties of solid–solid interfaces in high temperature corrosion, laminates and strained layer semiconductor devices. I will give some examples of how control of solid–liquid interfaces is being exploited in processing of cement and ceramics and, finally, speculate on some possible future developments.

2. Grain Boundaries — Historical Development

2.1. *Early metallography*

As is often the case, technological development is driven by military requirements and our story begins with patterns produced on sword blades from the middle east dating from about 500 AD to the nineteenth century. Figure 1 shows the blade of a Persian sword made of so-called "damask" steel which is characterised by intricate swirling patterns on its surface

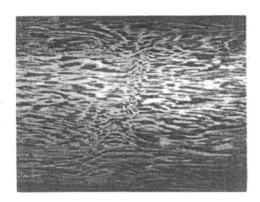

Fig. 1. Damask pattern etched on the blade of a Persian sword. (Ref. 24, reproduced with permission).

created by etching it in vinegar or mineral sulphate solution.[24] This pattern was prized not only for being decorative, but was also a "Quality Assurance" indicator because it was characteristic of the best material for combining toughness with retaining a cutting edge. The weapons were forged from high carbon steel imported from India and had a complicated internal structure as a result of a nonuniform distribution of carbon. The light parts in the etched pattern have a lot of carbon and are like cast iron (hard but brittle) whereas the dark parts are like low carbon steel (softer, but tough). The resulting composite structure gives the steel its unique and desirable properties and the chemical etching has revealed aspects of the internal structure of the steel which would not otherwise be visible. Subsequently this combination of mechanical polishing, chemical etching and optical observation (collectively known as metallography) became the favoured approach to studying the microstructure of materials.

However it was not known what gave rise to this internal structure. It is clear by looking at fractured surfaces of minerals that they have an internal structure composed of crystals, but the smooth swirling patterns on damask steel, and the malleability of most metals, do not give the impression of crystals. Nevertheless, Réaumur in 1724[24] was the first to recognise that metals were indeed formed from crystals and, with remarkable foresight, predicted that this was important in determining their properties; "Perhaps it will be found that it is on this shape of the grains, and their arrangement, that the ductility of metals, and of some other materials, depends." He was able to see these crystals clearly in specimens of antimony and lead that were fractured after being cooled from the molten state in different ways. Studies of meteorites by Schreibers and von Widmannstäten in 1813 first demonstrated that the crystalline structure of metals could be revealed by metallography (Fig. 2). The meteorite contains mainly iron with a little nickel. The microstructure is clearly crystalline with preferred directions of crystal growth and on a coarse enough scale to be directly printed onto paper (like a wood-cut). This type of structure is often found when one type of crystal

Fig. 2. An imprint from the polished and etched cross section through a meteorite produced in 1813. (Ref. 24, reproduced with permission).

transforms into another on cooling. It is coarse enough to be easily visible in the meteorite because it cooled very slowly.

To see the fine crystalline microstructure in technological materials such as steel requires a microscope; and the combination of metallography with microscopy, and subsequently photography, was pioneered by Henry Sorby in Sheffield. He was a keen amateur with an interest in the local steel industry who applied skills acquired in the study of geological specimens to metals and produced the first photomicrographs of metals in 1863.[12] He was moved to write the following verse, which reflects some of the persistence required for success, to describe his activities:

"Hard crystals I anatomise
And iron and steel too
And stubborn rock I shave and grind
And look them through and through."

2.2. The twentieth century

By employing these techniques the individual grains in materials could then be clearly seen. Although many people suspected the grains to be crystals, it was not clear how one grain differed from another and how they were joined together at the grain boundaries. Explaining these issues in terms of atomic properties has been the thrust of twentieth century work. In 1913, William and Lawrence Bragg were able to show by X-ray diffraction that even in metals the atoms were arranged in a regular crystalline structure and thus the grains were proved beyond doubt to be crystals in different orientations. But the nature of the grain boundaries was still obscure and needed the concept of the dislocation to reveal how the atoms in the two crystals forming the boundary could fit together. The dislocation (Fig. 3) was originally conceived to explain how metals manage to deform at unexpectedly low stresses but could not be observed directly at that time.

Nevertheless their structure and behaviour were simulated at Cambridge by Bragg and Nye[6] in some particularly elegant experiments using rafts of soap bubbles (the forces between the soap bubbles have similar characteristics to the forces between atoms in a metal). The bubble rafts were also used to simulate grain boundaries.[17] Figure 4 shows a boundary between two

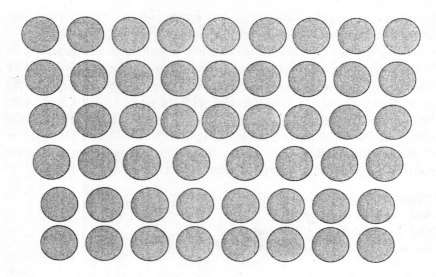

Fig. 3. Schematic view of an edge dislocation in a simple cubic crystal lattice. The core of the dislocation is where the extra plane of atoms terminates and the dislocation line is at the core of the dislocation perpendicular to the plane of the paper.

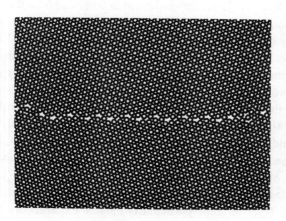

Fig. 4. A "grain boundary" produced by bringing together two rafts of soap bubbles tilted with respect to each other by an angle of 25° symmetrically about the line of their joining (Ref. 17).

"crystals" of bubbles, one of which is tilted with respect to the other. In the boundary is an array of dislocations which allow the bubbles to fit together quite well at the grain boundary. The bubble simulation also shows that there is some extra space at the core of each dislocation and it turns out that this extra space is important in controlling the properties of grain boundaries. With the development of the electron microscope it became possible to see inside metals at high magnification and at Oxford in Hirsch, Horne and Whelan[14] were able to observe directly the predicted arrays of dislocations at grain boundaries in aluminium.

Today, high resolution electron microscopy and computer simulations of the behaviour of atoms have become powerful techniques for the study of internal interfaces in solids. The computer simulations are used to predict the arrangement of atoms at an interface and the modern electron microscope is now capable of imaging them with atomic scale resolution.

3. Oxide Grain Boundaries and High Temperature Corrosion of Metals

In high temperature corrosion of a metal an oxide layer forms by chemical reaction between the metal and oxygen and this layer controls the corrosion rate because the metal has to pass through it (in the form of ions) to continue to react with oxygen. Figure 5 illustrates this for the oxidation of nickel to

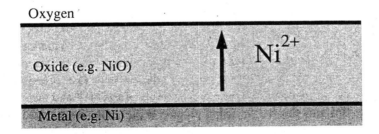

Fig. 5. Schematic illustration showing how growth of a layer of nickel oxide is controlled by nickel ions diffusing through it during oxidation of nickel.

nickel oxide. This is a particularly simple example that has been investigated in considerable detail, but manages to encompass the key features of the corrosion of technologically important high temperature alloys used in chemical engineering and power generation industries. Theories developed by Wagner over 60 years ago[26] enable the growth rate of the oxide to be predicted from independent measurements of the diffusion of nickel ions in nickel oxide. However, when this is done the predicted oxidation rate is found to be many orders of magnitude slower than is actually observed. The reason for this is that the nickel oxide layer formed by oxidation contains lots of grain boundaries that provide pathways which enable the nickel ions to diffuse through the oxide layer much more quickly than expected[5] and computer simulation has enabled us to understand how this occurs.

Figure 6 shows the structure of a grain boundary in nickel oxide from computer simulations by Duffy and Tasker.[9] The boundary has an array of open channels, similar to the ones that were produced by the bubble raft model, and these have been observed by high resolution electron microscopy (Fig. 7, Ref. 20). Diffusion of nickel ions along these channels is much quicker than diffusion through the crystal grains on either side. In the computer simulations it is possible to examine how this rapid diffusion takes place. It occurs by way of missing ions (vacancies) on some of the sites marked "A" and "C" in Fig. 6 (vacant sites are required for any diffusion of nickel ions to occur in nickel oxide crystals).

The computer simulations not only provide insight into the atomic movements responsible for corrosion being more rapid than expected, but also enable strategies for overcoming this problem to be explored. The simulations show that if cerium ions are introduced into the nickel oxide they segregate to particular sites on the grain boundary channels (marked "B" in Fig. 6) and they prevent the jump of nickel ions by trapping the vacancies on the "A" and "C" sites that are the cause of fast diffusion. (The "B" sites are on planes above and below those of the "A" and "C" sites in Fig. 6.) In this way the fast diffusion process is prevented by the cerium ions and the corrosion rate is greatly reduced. Other atoms (known as reactive elements) such as yttrium and lanthanum have been found to have the same effect. This means that if we put such elements into an alloy its corrosion

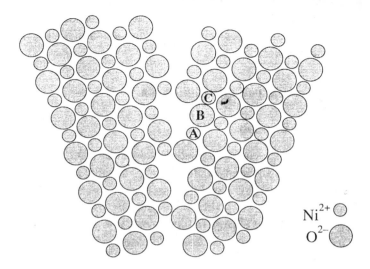

Fig. 6. Computer simulation of the structure of a grain boundary in nickel oxide in which two crystals have been tilted symmetrically by 36.9° about an axis perpendicular to the paper. In the layer of ions below and above the one shown the nickel and oxygen ion positions are interchanged. The labelled ion positions are the ones involved in the diffusion of nickel along the boundary perpendicular to the plane of the paper (Ref. 9).

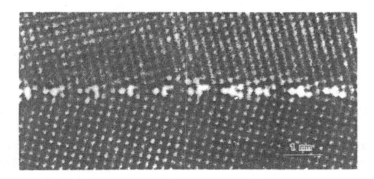

Fig. 7. High resolution transmission electron micrograph of a 26° symmetrical tilt boundary in nickel oxide in which the rows of ions show the same structure as predicted by the computer simulations (Ref. 20).

rate becomes very low. It works for metals other than nickel and is exploited widely in alloys that are particularly resistant to high temperature corrosion, such as "Fecralloy" steel.

4. Interfaces between Two Different Solids

4.1. *Oxide layers on metals*

Interfaces between oxide layers and bulk metals occur on components coated with oxides and on components undergoing high temperature corrosion, as in the example described in the previous section. In either case the adhesion at the interface between the oxide layer and the metal component is of crucial importance. In high temperature corrosion it is the oxide layer that protects the metal and prevents it being corroded very rapidly in a hostile environment. Therefore, if the oxide layer becomes detached from the metal, that protection will be lost. Because we now have an interface between two different materials the oxide layer is usually under a state of stress (the metal component to which it is attached is subject to a much smaller stress because it is usually much thicker than the oxide layer). A common cause of such stress is the difference in thermal expansion coefficients of the oxide and the metal. Oxide layers formed by a coating process or by oxidation are produced at high temperature and they usually have lower thermal expansion coefficients than the metal component on which they are formed. When they are cooled the metal will try to contract more than the oxide, but since the two are bonded together this creates a compressive stress in the oxide. Moreover, additional external stresses would also be applied to the oxide as a result of the use of the metal component in an engineering application. Therefore it is important to be able to measure the adhesion of the oxide layer to the metal substrate and to understand what determines this.

Figure 8 shows experimental measurements of the apparent adhesion between two metal substrates and the oxide layers formed on them by high temperature corrosion.[1] The measured stress is "apparent adhesion" rather than "true adhesion" because it is the external tensile stress applied perpendicular to the oxide surface that is required to pull it off the metal.

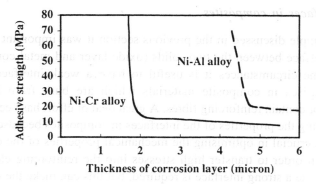

Fig. 8. The measured strength of the interface between nickel oxide corrosion layers and dilute nickel alloys showing how the interface becomes weak when the layer exceeds a critical thickness (Ref. 1).

It does not include explicitly the residual compressive stress which was generated by thermal expansion mismatch and is parallel to the oxide surface. The measurements show that there is a critical thickness of the oxide layer (different for different alloys) above which its apparent adhesion becomes very low. This critical thickness occurs when the energy stored in the oxide layer (which increases in proportion to its thickness) as a result of the thermal expansion mismatch stress becomes larger than the energy required to separate the interface between oxide and metal. The practical consequence of the existence of a critical layer thickness is that good adhesion between the oxide layer and the metal substrate can be achieved if the corrosion rate is sufficiently slow so that the oxide layer is always less than the critical thickness. Thus the same reactive elements that are used to slow down the corrosion rate have a double benefit; the thinner corrosion layers that they produce are extremely adherent, which is why they perform so well in practice.

The factors that determine the true adhesion of an interface between an oxide and a metal are the subjects of current research using a powerful combination of high resolution electron microscopy, computer simulation and mechanical measurements.

4.2. *Interfaces in composites*

In the example discussed in the previous section it was important to have a strong interface between the two solids (oxide layer and metal component), but in some circumstances it is useful to have a weak interface. Such a situation arises in composite materials which are built from laminated materials or contain reinforcing fibres. A great deal of effort has been devoted to controlling the properties of the interfaces in composites because interface strength is crucial in optimising the mechanical properties of the composite material. In order to transfer high stresses into the reinforcing elements of the composite a strong interface is required, but this can make the composite too brittle.

This is illustrated in Fig. 9 for a laminated composite in which the principal lamellae are brittle. Figure 9(a) shows what happens during fracture if the interface between the lamellae is strong. A crack, once initiated in one of the brittle lamellae, will readily propagate from one lamella to its neighbour resulting in a stiff but brittle composite. On the other hand, if the interfaces between the lamellae are weak the crack will be deflected along the interface [Fig. 9(b)] and the composite will not fail catastrophically. This makes the composite tougher, but not as stiff because weak interfaces are usually more compliant. Thus a suitable interfacial strength is required to optimise the combination of stiffness and toughness of the composite.

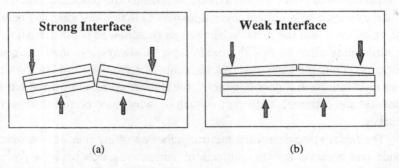

Fig. 9. Schematic illustration of the failure of brittle laminates in bending. In (a) the interface between layers is strong and failure is catastrophic, whereas in (b) the interface is weak and delaminates causing gradual failure.

An everyday example of this principle is the laminated glass used for car windscreens in which layers of brittle glass are joined together using a much weaker organic polymer. Similar concepts are now being investigated to fabricate laminated composites for high temperature applications in which layers of strong, stiff but brittle silicon carbide ceramic are joined together with a relatively weak graphitic interfacial bond.[8] In this way the toughness of the laminated structure is about 5 times that of silicon carbide on its own.

The principle is also well developed in nature in bones and shells. The main material in shells is calcium carbonate, which is hard and protective, but inherently brittle. To overcome this problem nature has devised a laminated structure in which the brittle calcium carbonate layers are joined with relatively weak proteins to give a tougher composite shell (Fig. 10). There is now a lot of interest in trying to make tough, stiff and lightweight composites by mimicking these natural materials.

Fig. 10. The structure of a natural shell in which the interface between brittle calcium carbonate laters delaminates during failure to create a tough composite from a brittle matrix (Ref. 7).

4.3. *Epitaxial interfaces between semiconductors*

Sometimes the atoms in different materials have the same crystalline arrangement. Such crystals can join together very well to form what are called epitaxial interfaces. If the spacing in one crystal is a little bit different from the other the crystals can fit together in two fundamentally different ways as illustrated in Figure 11. In one [Fig. 11(a)], the crystals deform so that the atoms line up in the two crystals at the interface. This can only happen if one of the crystals is a thin layer, in which case the layer has an internal stress parallel to the interface which distorts the crystal lattice to make it fit the lattice of the much thicker substrate crystal. This is similar to the stress discussed in Sec. 3.1 which arose from a thermal expansion difference between an oxide layer and a metal component. These uniformly strained epitaxial layers are referred to as "pseudomorphic". Alternatively as illustrated in [Fig. 11(b)], the crystals can be without overall strain (referred to as being relaxed), but at the interface the mismatch between the atoms appears as misfit dislocations, in a similar way to the dislocations at

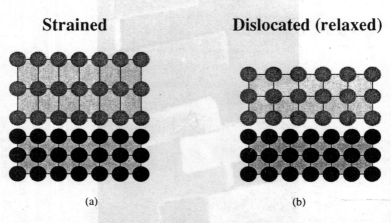

Strained	Dislocated (relaxed)
(a)	(b)

Fig. 11. Two configurations for the interface between two materials having the same crystal structure but different lattice parameters. The lattice parameter for the top crystal is larger than that of the bottom one. In (a) the top crystal has been strained uniformly to make it fit the bottom one. In (b) the crystals are not strained uniformly but a misfit dislocation is present at the interface (compare with Fig. 3).

a grain boundary discussed earlier. The relaxed configuration can be simulated by an interface between two rafts of soap bubbles of slightly different size (Fig. 12) and the arrays of misfit interfacial dislocations can be observed in real crystals by electron microscopy (Fig. 13).

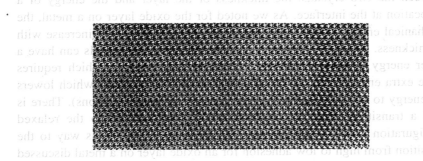

Fig. 12. An interface between two bubble rafts of different bubble size. The bubbles in the top raft are 10% larger than those in the bottom one and misfit dislocations are present at the interface (Ref. 19).

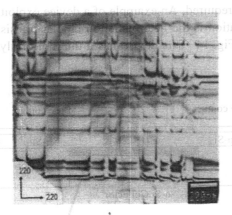

Fig. 13. Transmission electron micrograph showing networks of misfit dislocations (dark lines) at the interface between a 270 nm thick layer of $Si_{0.78}Ge_{0.22}$ on a Si (001) crystal substrate. The interface is viewed from above in this micrograph and the network of dislocations reflects the symmetry of the crystals (courtesy of A. Staton-Bevan).

The factors that determine whether the interface adopts the strained pseudomorphic or the relaxed dislocated structure were first clarified by Frank and van der Merwe[10] and have since been extended to more complicated situations.[2,3] The controlling factors are: the mismatch in lattice parameter between the two crystals, the thickness of the layer and the energy of a dislocation at the interface. As we noted for the oxide layer on a metal, the mechanical energy stored in the uniformly strained layer will increase with its thickness. When this becomes sufficiently large the crystals can have a lower energy if disloca-tions are created at the interface (which requires some extra energy) and the pseudomorphic stress is relaxed (which lowers the energy to compensate for the extra energy of the dislocations). There is thus a transition from the pseudomorphic configuration to the relaxed configuration above a critical layer thickness in an analogous way to the transition from high to low adhesion for an oxide layer on a metal discussed earlier.

Both strained and relaxed configurations are important in various new semiconductor device structures and the success of these devices depends on our ability to understand and control the formation of the two configurations as required. An example of a device exploiting the uniformly strained configuration is the heterojunction bipolar transistor (HBT) shown schematically in Fig. 14. This device has a thin uniformly strained layer of

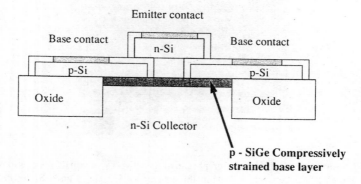

Fig. 14. The structure of a heterojunction bipolar transistor (HBT) in which the base is a thin layer of strained silicon–germanium alloy grown epitaxially on the silicon substrate (Ref. 21).

silicon–germanium alloy, grown pseudomorphically on silicon, acting as the base of the transistor. Both silicon and germanium have the same crystal structure and can easily mix to form alloys, but the germanium atoms are larger than the silicon atoms and so the silicon–germanium alloy has a larger crystal lattice size than pure silicon. As a result, the pseudomorphic alloy layer must be compressed to make it fit the silicon substrate. The compression changes the electrical properties of the alloy in a beneficial way and increases the moblity of electrical charge carriers (in this case electron holes) so that transistors that can switch more quickly can be made. For good electrical performance there must be no dislocations at the interface between the alloy and the silicon substrate and so the alloy layer must be thinner than its critical thickness. In the quest for new electronic devices (e.g. for faster computers and telecommunications), speed is a major goal and these HBTs are the fastest silicon devices with switching frequencies approaching 200 GHz (Fig. 15) which is almost three times faster than the conventional silicon bipolar transistor.

Even faster speeds are expected from another device called the heterojunction field effect transistor (HFET) which has the layered structure

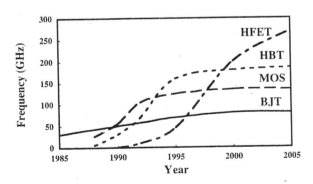

Fig. 15. Improvement in the switching frequency of silicon-based semiconductor devices with time showing expected trends to 2005. The bipolar junction transistor (BJT) and metal oxide semiconductor (MOS) field effect transistor are conventional silicon devices whereas the HBT and HFET contain silicon–gemanium alloy layers grown on silicon (Ref. 16).

shown in Fig. 16, Ref. 23. In this device the idea is to produce a thin layer of silicon uniformly strained in tension. This can be grown pseudomorphically on a silicon–germanium alloy substrate because the alloy, having the larger lattice' constant, will stretch the silicon layer to make it fit. As with the HBT, this layer must be free of dislocations and therefore thinner than its critical thickness. However, it is not possible to obtain good quality bulk silicon–germanium alloys to use as the substrate and so the alloy is itself grown as a layer (called a virtual substrate) on a bulk silicon single crystal substrate. To obtain the true lattice parameter in the alloy layer it must be fully relaxed and therefore much thicker than its critical thickness. To make sure the misfit dislocations remain in the alloy and do not extend into the stretched silicon layer the germanium content in the alloy is increased gradually with distance above the silicon substrate. These devices represent a high degree of refinement in the area of "interface engineering" and the HFET is expected to reach speeds approaching 300 GHz (double that of conventional silicon field effect transistors).

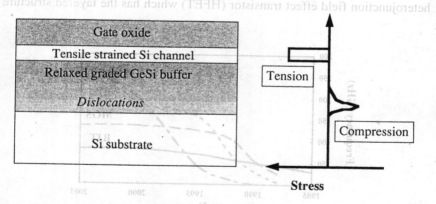

Fig. 16. The structure of the gate region in a heterojunction field effect transistor (HFET). The germanium concentration gradually increases with distance from the silicon substrate in the silicon–germanium alloy buffer layer. This is much thicker than its critical thickness and so is relaxed. The silicon layer is thinner than its critical thickness and is uniformly strained in tension. The diagram on the right shows how the stress varies with position in the structure.

5. Interfaces between Solids and Liquids

5.1. *Processing ceramics*

Modern engineering ceramics are made using very fine powders comprising particles less than one micron in size. These tiny ceramic particles are very difficult to handle when dry and will try and stick together, even if mixed with water. It is possible to prevent them from sticking by attaching electrical charges, or organic polymer molecules to their surfaces which then repel one particle from another (Fig. 17). These principles have been used in ceramic processing for many decades, but in recent years they have been the subject of increased activity as the emphasis has shifted to smaller particle sizes and more reliable products. By controlling the interface between the ceramic particles and the liquid they are dispersed in it is now possible to produce dispersions of fine ceramic powders that are sufficiently fluid to be poured into moulds yet contain over 60% by volume of solid material. In addition it is possible to then change the interface properties so that the particles will stick back together again to form a solid shape or gel. This is exploited in gel casting processes to make bulk ceramic shapes[11] and the sol-gel processes for making ceramic fibres, powders and coatings.[4]

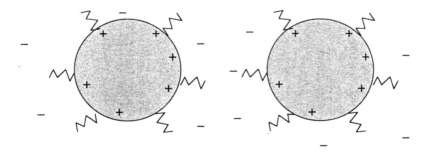

Fig. 17. Schematic illustration showing how small colloidal solid particles suspended in a liquid can be prevented from sticking together by the adsorption of electrical charge or polymer molecules at the solid–liquid interface.

5.2. *High strength cements*

Organic molecules called superplasticisers are used widely in the concrete construction industry. They are added to the cement when it is mixed with water and they adsorb onto the surfaces of the cement particles giving them an electric charge to make them repel each other and produce a more fluid concrete mixture. In conventional concrete this enables less water to be used for mixing the concrete and results in stronger and more durable concrete after it sets and cures.

The same principle has been used to increase the strength of special cements even further. Cement particles are rather large (e.g. 20 microns) by comparison with the much finer particles used for modern engineering ceramics. Therefore it is possible to fill in the spaces between the cement particles with finer particles to give a cement "densified using small particles" or DSP.[18] Fumed silica is a commercial product that is suitable for this as it is not expensive and is chemically compatible with cement, which is mainly calcium silicate. Because the silica particles in the fumed silica are very small, this cement can only be mixed using a superplasticiser to absorb onto the silica particles and prevent them from sticking together. The cured cement has a strength about four times that of conventional cement (Fig. 18)

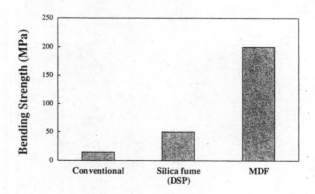

Fig. 18. The bending strength of conventional Portland cement compared with those of cements that have been modified with colloidal silica particles and polymers. MDF stands for "macro-defect-free".

and is used for special cement products such as high strength tiles. Even higher strengths are possible in cementitious materials. "Macro-defect-free" (MDF) cement[15] is a composite material of cement and polymer in which the polymer increases the cement strength by acting as a super-plasticiser during processing, filling in space between cement particles and then hardening (crosslinking) as the cement cures. MDF is not shaped as a fluid slurry in the same way as a conventional cement, but as a pliable "dough". This enables large voids such as air bubbles to be removed so that they cannot lead to defects in the final material which, if present, would lower its strength. MDF cements can achieve strengths over ten times that of conventional cement (Fig. 18) but have not been exploited widely to date because the polymers that have been used, have not had sufficient long term stability.

6. Future Opportunities and Concluding Remarks

Some of the likely future directions have already been mentioned in the previous sections. Computer simulation and observations of atomic structure are expected to be applied to more complicated systems as computational power and experimental techniques for structural characterisation both improve. As new ways are also found to build controlled interfaces by appropriate materials processing methods, "interface engineering" will expand to embrace an ever widening range of materials.

We can also expect to see application of graded interfaces in a much wider range of materials and structures. Graded interfaces avoid many of the mechanical problems that arise from the abrupt change from one material to another that occurs at a "normal" interface. They are already being investigated in a range of new technologies, ranging from protective coatings to electronic devices, and in joints between different bulk materials. These graded structures will find even wider application as cheaper techniques to produce them are developed.

Multilayer structures are also becoming more widely used and this is another trend that is set to continue. Applications span all size scales from nanometres to millimetres. Individual semiconductor layers in multilayer electronic and opto-electronic devices can be as thin as 1 nanometre.

Multilayer coatings of alternating hard and soft materials with individual layers in the region of 0.1 micron thickness are being developed for wear resistance. Multilayer electrical capacitors, actuators, integrated circuits and other microelectronic devices have layer thicknesses between 0.1 and 10 microns. Finally, laminated devices incorporating ceramic membranes (for example, solid oxide fuel cells for electricity generation and gas separators) are being developed with layer thicknesses in the range 10 microns. Just one element of these ceramic membrane devices typically has within it ten interfaces between different materials, including metals, ceramics and glass, all of which must work together in terms of mechanical, chemical and electrical properties.

Nanocrystalline materials and nanocomposite materials are emerging classes of new materials with some unique properties. In these materials as many as half of the atoms are located in grain boundaries and interfaces rather than in the crystal lattice. Thus all their properties are dominated by the internal interfaces within them. One particularly interesting type of nanocomposite material is the so-called organic–inorganic hybrid polymer. These can take many different forms[22] and the structure of such a hybrid polymer is illustrated schematically in Fig. 19. The structure contains inorganic polymeric chains (in this case of $-Si-O-Si-$ linkages) and organic polymeric chains (in this case a polyacrylate) and the two chains are crosslinked with organic units. Thus the whole material can be viewed as being made entirely of an interface between interpenetrating organic and inorganic polymers. The attraction of these hybrids is that they have properties that are intermediate between an inorganic glass and an organic polymer. They are harder, stiffer and more stable than organic polymers yet tougher than glass.

Interfaces that were once regarded as sources of weakness and of other problems in materials can now not only be controlled to avoid these problems, but are being viewed positively as offering opportunities for new materials and structures. The field has come a long way since it was first recognised that metals were crystalline and contained internal interfaces in the form of grain boundaries. The understanding, control and exploitation of interfacial properties in an ever widening array of materials will surely continue to be

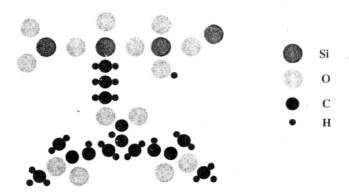

Fig. 19. An example of an organic–inorganic hybrid polymer, or nanocomposite, in which an inorganic polymer network (silicate in this example) is crosslinked to an organic polymer network (methacrylate in this example).

key aspects of materials science and engineering in the future. It can reasonably be claimed that we have both beaten them and joined them.

Acknowledgements

I would like to thank all the colleagues with whom I have worked in the past at Leeds University and AEA Technology (Harwell), and those with whom I am currently interacting at Imperial College, for their help and companionship in the research we have carried out together.

References

1. A. Atkinson and R.M. Guppy, *Mat. Sci. Technol.* **7**, 1031, 1991.

2. A. Atkinson and S.C. Jain, *J. Phys.: Condensed Matter*, **5**, 4595, 1993.

3. A. Atkinson, K. Pinardi and S.C. Jain, *Semiconductor Sci. Technol.* **11**, 1271, 1996.

4. A. Atkinson and D.L. Segal, *Ceramic Technology International*, 187, 1993.

5.   A. Atkinson, R.I. Taylor and A.E. Hughes, *Phil. Mag.* **A45**, 823, 1982.

6.   W.L. Bragg and J.F. Nye, *Proc. Roy. Soc.* **A190**, 474, 1947.

7.   P. Calvert, *Mat. Res. Soc. Bull.* **37**, October 1992.

8.   W.J. Clegg, K. Kendall, N. Alford, T.W. Button and J.D. Birchall, *Nature*, **347**, 455, 1990.

9.   D. Duffy and P.W. Tasker, *Phil. Mag.* **A47**, 817, 1983.

10. F.C. Frank and J.H. van der Merwe, *Proc. Roy. Soc. London*, **A198**, 216, 1949.

11. T.J. Graule, W. Si, F.H. Baader and L.J. Gauckler, *Ceramic Transactions*, **51**, 457, 1995.

12. C. Hammond, *Microscopy*, **35**, 581, 1987.

13. P.B. Hirsch, R.W. Horne and M.J. Whelan, *Phil. Mag. Ser. 8*, **1**, 677, 1956.

14. K. Kendall, A.J. Howard and J.D. Birchall, *Philos. Trans. Roy. Soc. London Series*, **A310**, 139, 1983.

15. U. König and J. Hersener, *Proc. of Gettering and Defect Engineering in Semiconductor Technology Conference*, eds. H. Richter, M. Kittler and C. Claeys, *Solid State Phenomena*, **47–48**, SCITEC Publications, Switzerland, 1996, 503.

16. M.W. Lomer and J.F. Nye, *Proc. Roy. Soc.* **A212**, 576, 1952.

17. P. Lu, G. Sun and F.J. Young, *J. Amer. Ceram. Soc.* **76**, 1003, 1993.

18. J.W. Matthews, *J. Vac. Sci. Technol.* **12**, 126, 1975.

19. K. Merkle and D.J. Smith, *Ultramicroscopy*, **22**, 57, 1987.

20. D.V. Morgan and R.H. Williams, *Physics and Technology of Heterojunction Devices*, Peter Peregrinus Ltd., London, 1991, 201.

21. B.M. Novak, *Advanced Materials*, **5**, 422, 1993.

22. A. Sadek, K. Ismael, M.A. Armstrong, D.A. Antoniadis and F. Stern, *IEEE Trans. Electron Devices*, **43**, 1224, 1996.

23. C.S. Smith, *A History of Metallography* University of Chicago Press, 1960.

24. F.L. Vogel, W.G. Pfann, H.E. Corey and E.E. Thomas, *Phys. Rev.* **90**, 489, 1953.

25. C. Wagner, *Z. Phys. Chem.* **B21**, 25, 1933.

Professor Rees D. Rawlings

Rees Rawlings obtained a BSc (Eng) in Metallurgy at Imperial College in 1964 and stayed on in the Department of Materials (then called the Department of Metallurgy) to study for a PhD. He joined the staff in 1966 and was awarded his PhD in 1967. His research field is the correlation of microstructure and properties, mainly mechanical properties but in some cases functional properties such as electrical and magnetic. Early work was mainly on brittle metals and intermetallic compounds but an interest in ceramics, including glasses and glass-ceramics, also soon developed. This work has recently been extended to composites and functionally graded materials based on metals and ceramics.

He is a Fellow of the Institute of Materials, was awarded a DSc in 1989 and the Pfeil Medal of the Institute of Materials in 1990. He is Chairman of the Steering Committee of the Centre for Composite Materials, Pro Rector (Educational Quality) and Deputy Editor of Journal of Materials Science and Journal of Materials Science Letters.

BRITTLENESS — A TOUGH PROBLEM

REES D. RAWLINGS

Department of Materials
Imperial College of Science
Technology and Medicine
Prince Consort Road
London, SW7 2BP, UK
E-mail: r.rawlings@ic.ac.uk

1. Introduction

We learn early in life the hard lesson that a pottery, china or glass mug shatters into numerous pieces when dropped onto a hard floor whereas their metal counterparts produce a satisfying clang and remain intact, although perhaps dented. Pottery and china are examples of the class of materials known as ceramics. Ceramics and glasses, as we know from experience, fail in a catastrophic manner — in other words they are brittle. In contrast metals are generally difficult to break and are therefore said to be tough. In simple terms, toughness is the opposite of brittleness. However, metals are readily dented or deformed and such permanent changes in shape is the consequence of the phenomenon of plastic deformation.

Now, the fact that ceramics are brittle does not mean that they cannot carry heavy loads. For example a ceramic mug may be able to carry an elephant weighing over 1000 kg [Fig. 1(a)]. I do not recommend trying to place an elephant on a mug as, besides the obvious practical difficulties, the results are rather uncertain. If we had succeeded in placing the elephant on

(a)

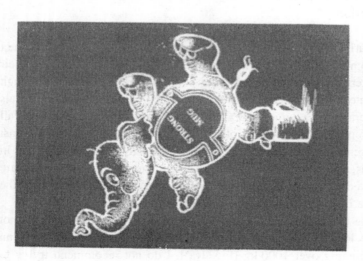

(b)

Fig. 1. Elephant balancing on a ceramic mug (a) strong mug (small flaws) (b) weak mug (large flaws).

a mug and then attempted to transfer carefully the elephant to another, apparently identical mug, the latter may well have broken with disastrous consequences [Fig. 1(b)]! The conclusion we are forced to draw is that ceramics are brittle and that their strength is variable — sometimes they will perform satisfactorily other times not. Why is this?

To begin to understand the mechanical behaviour of ceramics we must first look at the effect of notches, or flaws, on load carrying capacity. Consider the two ceramic samples illustrated in Fig. 2; one has smooth sides while the other has a notch on each side. The important point to note is that the distance between the notches is exactly the same as the width, *w*, of the unnotched sample. Let us pull on the samples with an ever increasing force *P*. The notched sample fails first, i.e. fails at a lower applied force *P*, a fact that demonstrates that notches, or flaws, have a detrimental effect on the strength of brittle materials. The different strengths of the two ceramic mugs can therefore be attributed to the size of the flaws; the strong mug had small flaws and the weak mug had large flaws.

To investigate further the dependence of strength on flaw size we load a notched specimen made, not from a ceramic, but from a photoelastic polymer. The use of a photoelastic polymer allows us to identify the regions of high stress that develop in the specimen when a force is applied (stress is force divided by the area over which it is acting). The highly stressed regions in a photoelastic materials are covered with closely spaced black

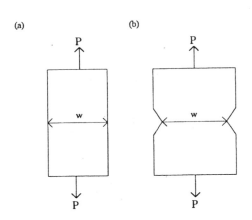

Fig. 2. Effect of a notch. (a) unnotched and (b) notched specimens of a ceramic being pulled by an ever increasing force *P*. The notched specimen fails first.

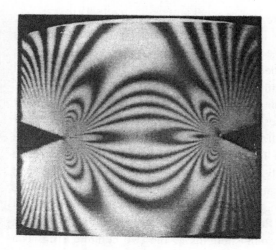

Fig. 3. A loaded notched photo-elastic specimen showing the regions of high stress at the notch tips.[1]

and white fringes when viewed with transmitted polarised light. In this experiment the fringes first appear at the notch tips and, as the applied force increases, extend to cover the whole specimen but with the fringe spacing being finer at the notch tips (Fig. 3). From this we conclude that notches (flaws) act as stress concentrators, that is the stress at the tip of a flaw is higher than in regions remote from the flaw.

So at this point we have learnt that flaws, which will always be present in materials, concentrate stress [Fig. 4(a)] and lead to brittle fracture of ceramics [Fig. 4(b)]. However, metals will also have stress concentrating flaws yet do not usually fail in a brittle manner. The reason for this is that the high stress at the flaw tip in a metal causes plastic deformation which blunts the flaw and reduces its ability to concentrate stress [Fig. 4(c)].

To explain why these two classes of materials differ in their response to the stress present at a flaw tip we have to look in depth at the structure of materials. When we are working at the microstructural level, we are studying features of typically a fraction of a micron to several hundreds of a micron. The microstructure of Fig. 5 has two constituents, known as phases, and is therefore said to be a two-phase microstructure. We also need to look at materials on even a finer scale, namely the atomic level, in order to determine

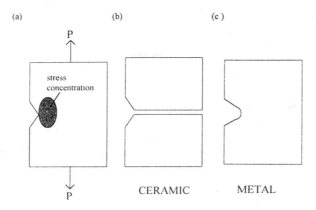

Fig. 4. Effect of flaws (a) stress concentration (b) brittle failure of a ceramic or glass (c) blunting of flaw in a metal by plastic deformation.

Fig. 5. The structure of materials — microstructural and atomic levels.

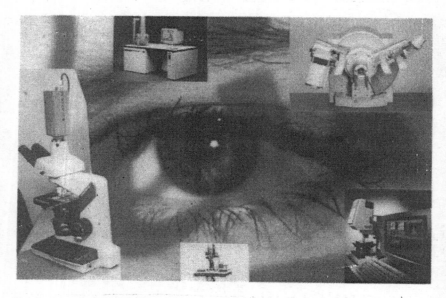

Fig. 6. Some of the techniques in common use for investigating the structure of materials (clockwise from top left: scanning electron microscope, X-ray diffractometer, image analyser, transmission electron microscope, light microscope).

Fig. 7. Cartoon (courtesy Laboratory Digest).

how the atoms are arranged. At the atomic level we are working on the nanometre scale where a nanometre is one-millionth of a millimetre.

To obtain this structural information materials scientists use a variety of techniques, some of which are illustrated in Fig. 6. These are standard techniques that are used on a daily basis by materials scientists to delve into the structure of materials. We think there is nothing odd in wanting to investigate the fine structure of materials in this way but the general public might think this behaviour is far from normal (Fig. 7)!

2. Deformation of Metals

The difference in behaviour of ceramics and metals is readily demonstrated by carrying out mechanical tests such as the tensile test illustrated in Fig. 2(a). The results from a tensile tests are given as graphs of applied stress against the strain (change in length divided by original length) produced. The stress–strain curve for a ceramic is linear up to the fracture stress where catastrophic failure occurs (Fig. 8). In the linear region the deformation, which is reversible and instantaneous, is termed elastic deformation. The stress–strain curve for a metal also has an initial linear region where the deformation is elastic, but this is usually followed by an extensive nonlinear region which corresponds to permanent or plastic deformation (Fig. 8). The total work needed to break a material, known as the work of fracture and measured in energy per unit volume, is the area under the stress–strain curve (shaded area on curves of Fig. 8); the greater the work of fracture the tougher is the material and thus this criterion confirms the superior toughness of metals.

Let us turn our attention to metals and try to understand why they deform plastically so readily. If we plastically deform a metal sample that has been polished previously to a mirror finish such that the surface appears featureless when viewed by a light microscope and again view the surface we will see a considerable change. As a consequence of the plastic deformation the surface will be covered in fine lines, called slip lines. Closer observation of these lines will reveal that they are in fact fine steps on the surface (Fig. 9). Such observations led to the early model for plastic deformation, known as the block slip model, whereby blocks of the crystal

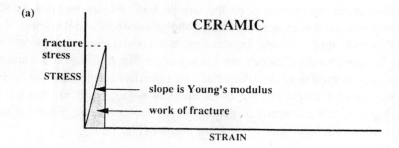

Fig. 8. Comparison of the stress–strain curves from (a) a ceramic and (b) a metal.

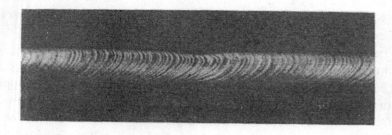

Fig. 9. Slip steps on the surface of a plastically deformed single crystal of a copper-aluminium. (This photograph was first published in Ref. 2 and was used by Sir Harold Carpeter and J.M. Robertson, who were metallurgists at Imperial College, in their classic two volume textbook[3]).

were considered to slip over each other on specific crystallographic planes and in specific crystallographic directions termed slip planes and slip directions respectively (Fig. 10). It should be noted that if plastic deformation occurs by this block slip mechanism all the bonds between the atoms

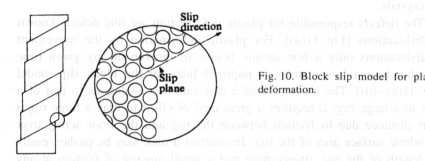

Fig. 10. Block slip model for plastic deformation.

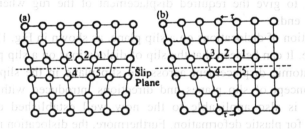

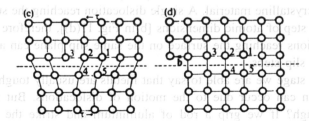

Fig. 11. Dislocations and plastic deformation (a) edge dislocation (b) stress applied and the dislocation moves slightly to the left (c) stress increases and the dislocation moves one atomic spacing to the left (d) this procedure is repeated until the dislocation leaves the crystal forming a step of height b.[4]

across the slip plane have to be broken at the same time — this requires a great deal of energy and calculations show the stress needed to be applied to the material is greater than that experimentally measured for plastic deformation. It follows the block slip model is not valid, instead it has been found that plastic deformation takes place by the movement of defects in the crystals.

The defects responsible for plastic deformation are line defects known as dislocations [Fig. 11(a)]. For plastic deformation by the movement of dislocations only a few atomic bonds are broken at any given time and therefore a lower stress is required than for the block slip model [Fig. 11(b)–(d)]. The movement of a dislocation is analogous to that of a ruck in a large rug. It requires a great deal of effort to drag a large rug a short distance due to friction between the rug and the floor acting over the whole surface area of the rug. In contrast a ruck may be pushed easily the length of the rug, overcoming just a small amount of friction at any given time, to give the required displacement of the rug when the ruck reaches the end.

The motion of dislocations on a slip plane, as shown in Fig. 11, is called slip or glide. It can be seen that the slip of dislocations on a slip plane gives a relative atomic displacement across the slip plane in the slip direction; thus the concept of slip planes and directions introduced with the block slip model is also applicable to the now well established dislocation mechanism for plastic deformation. Furthermore, the dislocation mechanism is consistent with the observation of slip steps on the surface of a plastically deformed crystalline material. A single dislocation reaching the surface will give a slip step of atomic dimensions [b in Fig. 11(d)], therefore thousands of dislocations reaching the surface on the same slip plane can account for the visible slip steps.

At this stage we are able to say that metals are usually tough as plastic deformation can occur due to the motion of dislocations. But are metals always tough? If we grip a rod of aluminium and strike the ground as hard as we can, we will not succeed in breaking the rod although it might bend. Repeating the experiment with a rod of copper would yield a similar result — both these metals are tough. However the situation is very different if we test a mixture of aluminium and copper in the proportions 2 atoms of copper to 3 atoms of aluminium. This material is extremely brittle and would shatter into many pieces. For some reason dislocation motion must be severely limited in the Cu–Al mixture thus restricting plastic deformation and the associated flaw blunting.

When we mix metals in particular proportions we may form what are known as intermetallic compounds or ordered solid solutions. Here are some well known examples — Al_3Cu_2, NiAl, $TiAl_3$, Fe_2Nb, FeCo — all have simple ratio of the constituent atoms. The characteristic of these materials is that the two constituent atoms are arranged in an ordered manner (Fig. 12) and it is this ordered arrangement that makes dislocation movement more difficult and results in brittleness.

A commercial ordered solid solution that we studied for many years at Imperial College is FeCo, which is a magnetic material that has to be produced in the form of thin sheet for the relevant industrial applications. When the FeCo is ordered it is brittle and cannot be fabricated into sheet and therefore a complex manufacturing procedure has to be followed (Fig. 13). At high temperatures (above about 725°C) the material is disordered and it can therefore be rolled at these temperature; this process being termed hot rolling. If the material is slowly cooled from the hot rolling temperature the atoms will rearrange into the ordered structure and the material will be brittle.

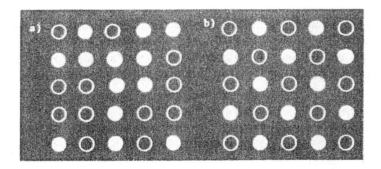

Fig. 12. A mixture of A and B atoms in a 1:1 ratio giving (a) a disordered solid solution (b) an ordered solid solution. An example of such a mixture is FeCo.

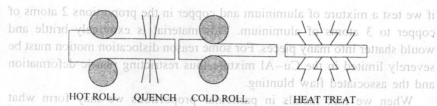

HOT ROLL   QUENCH   COLD ROLL   HEAT TREAT

Fig. 13. Production of FeCo soft magnetic alloy.

This is prevented by rapidly quenching from the hot rolling temperature — thus the atoms do not have time to rearrange and the disordered structure is retained. The disordered structure can of course be deformed and so the quenched material can be cold rolled to the required thickness and surface finish. In this state the magnetic properties are poor and the final stage in the processing is a heat treatment which recrystallises and orders the material.

The critical stage in the processing of FeCo is the quench from hot rolling. Whether the FeCo alloy is ordered, partially ordered or disordered after quenching depends on the efficiency of the quench and the presence of ternary alloying additions. Traditionally an addition of 2% vanadium has been used to facilitate retention of disorder. We found that less than 1% of tantalum or niobium was equally effective in this respect and at the same time gave better magnetic properties (a higher magnetic saturation) than the vanadium bearing alloy [e.g. 5–7]. Magnetic FeCo alloys with tantalum or niobium additions are now offered commercially by Telcon Limited, UK.

Metals do not have to be ordered to be brittle. Steel, which may be viewed as essentially an alloy of iron and carbon, may be heat treated to give markedly different properties. Consider two sample of the same composition but one has been slowly cooled (furnace cooled) from a high temperature and the other quenched from the same high temperature. The former would be tough whereas the latter would be very brittle requiring hardly any energy to break it. The reason for the difference in toughness is that the specimens have different microstructures (Fig. 14).

The slowly cooled material has a two-phase microstructure consisting of ferrite, which is almost pure iron and is present with granular and lamellar

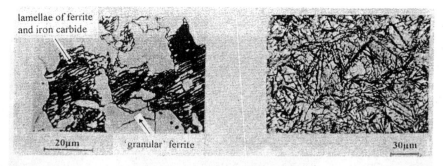

lamellae of ferrite
and iron carbide

20μm   'granular' ferrite   30μm

Fig. 14. Microstructure of steel (a) slowly cooled — two-phase microstructure (b) quenched — single-phase martensitic structure.

morphologies, and fine lamellae of iron carbide. Thus the carbon is concentrated in the iron carbide, whereas quenching gives a single-phase structure, known as martensite, in which the carbon is evenly dispersed as individual atoms throughout the iron. The carbon atoms occupy the interstices or holes between the iron atoms and are extremely effective in hindering the motion of dislocations — hence the brittleness of the quenched sample. The quench martensitic sample is also very hard and the higher the carbon content the harder and the more brittle is the steel.

At this point I would like to refer you the photographs of a necklace and a brooch presented in Fig. 15. These glittering, elegant antiques are not made using hundreds of diamond or other precious stones, but are fabricated of steel. This type of jewellery, known as cut-steel jewellery, was firmly established in Britain by the seventeenth century, expanded in the eighteenth and early nineteenth centuries and was effectively dead by the end of the nineteenth century. It was expensive jewellery and used in various forms by both sexes as shown by the cartoon of an eighteen century beau dazzling his lady friend with his cut-steel buttons (Fig. 16). The manufacture of cut-steel jewellery was a complex process that attained the required result by making full use of the dependence of properties on heat treatment and carbon content.[9]

Each jewel in a cut-steel article is either a facetted, approximately spherical bead, as used in the necklace, or rather like a mushroom with a

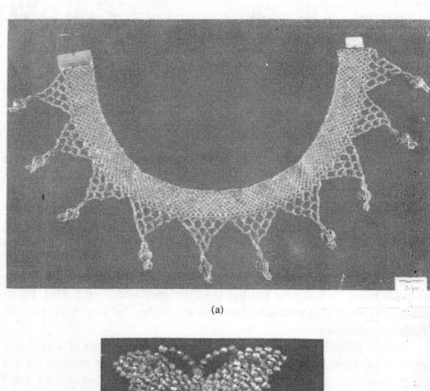

(a)

(b)

Fig. 15. Cut-steel jewellery (a) necklace made with beads (b) brooch made from "gems" such as those shown in Fig. 17.

Fig. 16. A gentleman in 1777 dazzles his lady friend with his cut-steel buttons.[8]

Fig. 17. A single "gem" from an article of cut-steel jewellery.[9]

facetted cap (Fig. 17). The shank of the latter is rivetted or screwed into a backplate and only the facetted cap or stud is seen as illustrated by the brooch. The microstructure of both the shank and the stud is martensitic, but cleverly, the shank is made from very low carbon steel while the stud is a higher carbon steel which, although more brittle, is also much harder and wear and corrosion resistant. This combination of steels was achieved by heat treating the jewels with the studs in contact with a carbon-donating material, such as bone-dust, while protecting the shanks from the bone-dust by surrounding them in clay. Consequently during the heat treatment the carbon content of the studs was increased and that of the shanks remained unchanged. The heat treatment was terminated by quenching in order to produce the martensitic structure. Of course the studs had to be highly polished and in an account of the manufacture of cut-steel published in London in 1830,[10] it is stated:

"Polishing is effected by means of putty, or the combined oxides of a mixture of lead and tin, finely levigated; and it is applied, mixed either with water, or, still better, in proof spirit, upon the palms of the hands of women, for a considerable length of time; indeed, until the fine black polish or lustre of hardened steel is at length produced.

No effectual substitute for the soft skin which is only to be found upon the delicate hands of women, has hitherto been met with."

I must admit I wonder just how soft hands would be after working 10 hours a day, six days a week polishing steel!

3. Ceramic Components and Fibres

Let us leave metals and return to ceramics, which we have learnt fail in a brittle manner from flaws within, or more commonly on the surface of, the component. Ceramics are crystalline materials and like metals have dislocations but the dislocations do not readily glide on the slip planes and also, due to the complex crystallographic structures of ceramics, there are fewer slip planes. Glasses are not crystalline and do not have a regular arrangement of the atoms, thus dislocations do not exist. Hence it is not possible for dislocation motion to relieve stress concentrations at the tips of flaws in ceramics and glasses.

This brittleness does not mean that ceramics are not good engineering materials, indeed they have many outstanding characteristics such as good wear and erosion resistance and the ability to operate at high temperatures, and are successfully employed in many applications. Satisfactory service performance requires: (1) control of the processing to give the required microstructure and to minimise flaw size and (2) good design.

A ceramic component that fulfils these requirements is the femoral head of a hip prosthesis (Fig. 18). The spherical head is made from aluminium oxide or zirconium oxide, which are more commonly known as alumina and zirconia respectively.[11-14] The ceramics have a fine microstructure and the spheres are produced with a smooth surface finish; these two features optimise the strength and wear resistance which is superior to that of their metallic rivals. Some typical mechanical property values for alumina and zirconia are presented in Table 1; also included in the table for comparison purposes are

Fig. 18. Hip prosthesis with a metal stem and a ceramic femoral head.

Table 1. Mechanical properties of bone and the ceramics used for femoral heads.[13–16]

	Young's Modulus (GPa)	Strength (MPa)		Toughness (MPa m$^{1/2}$)
		Bend	Compressive	
Cancellous Bone	0.02–1.70		0.15–5.0	
Cortical Bone	15–25	46–156	117–216	2.2–5.7
Alumina	380	500–560	4100	4.0–6.0
Zirconia	210	680–904	2000	9.4–10.7

data for cancellous and cortical bone. Tensile stresses are more likely to cause catastrophic failure than compressive stresses and account must be taken of this at the design stage of a ceramic component. The prosthesis is designed, with particular care taken with the means of attachment of the head to the metal stem, so as to minimise the tensile stresses in the ceramic head.

How else can we ameliorate the brittleness of ceramics and maximise their strength? I would like to start to answer this question by describing an experiment that was carried out by C.V. Boys at Imperial College over 100 years ago.[17] The equipment was very simple; it consisted of a crude crossbow fixed to a support and a gas jet for heating (Fig. 19). A short length of glass was attached between the arrow and the support for the bow and then heated with the gas flame. When the glass had reached an appropriately high temperature the arrow was fired and pulled a fine fibre in its wake from the hot glass. By this means Boys was able to produce glass fibres approaching 30 m in length and only a few microns in diameter. He also recognised the outstanding property of the fibres, namely their high strengths. Although he admitted to not having carried out any careful mechanical property measurements he was able to state that the strength was of the order of 790 MPa and that strength increased as the fibres became finer. We now know that the reasons for the high strength of a fibre are the smooth surface and the dimensions (small diameter) of the fibre limiting the flaw size. The flaw size being limited by the diameter of the fibre also accounts for the dependence of strength on the fineness of the fibre, i.e. the smaller the diameter, the smaller the flaw and the greater the strength.

(a)

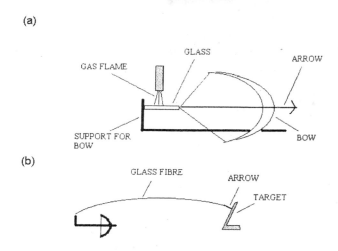

(b)

Fig. 19. Production of glass fibres by the method developed by Boys.[17]

Nowadays we manufacture glass fibres more efficiently than Boys was able to, and without needing crossbows! Molten glass is extruded under gravity from a melting tank through an orifice and rapidly pulled to draw it down to a fibre of about 10 μm diameter. Normally there are over 200 orifices used on the same melting tank and the fibres are drawn at speeds up to 50 m/s. Ceramics, such as silicon carbide and alumina, are also currently available in fibrous form but the processing technology is fibre specific and differs from that described for glass. Typical properties of modern glass fibres, and of representative natural and other synthetic fibres, are given in Table 2. The data of Table 2 demonstrate that we are able to manufacture glass and ceramic fibres with high strengths in spite of their inherent brittleness.

Another property of concern is the Young's modulus, which is defined as the slope of the initial linear region of the stress–strain curve (Fig. 8). The practical significance of Young's modulus is that it determines the reversible elastic strain resulting from a given stress when a material is in service. Usually the amount of elastic deformation of a component is small

Table 2. Properties of natural and synthetic fibres.[18]

Fibre	Density (Mg/m^3)	Young's Modulus (GPa)	Strength (MPa)
Natural Fibres			
Hemp			460
Jute	1.3	55.5	442
Sisal	0.7	17	530
Asbestos	2.56	160	3100
Synthetic Fibres			
Alumina	3.9	380	1600
Glass	2.54	70	2200
Carbon (IM)	1.76	290	3100
Silicon Carbide	2.4	280	2000
Aramid[@]	1.44	130	2900

[@] Aramid is an organic synthetic fibre.

and not noticeable; for example if you are sitting on a chair reading this book both the chair and the floor elastically distorts under your weight! Obviously we prefer materials to be stiff, that is to deform elastically as little as possible while in service, and therefore require high values for Young's modulus. The synthetic ceramic (alumina, carbon, silicon carbide) and natural (asbestos) inorganic fibres have high Young's moduli and hence high stiffnesses. Glass fibres do not compare well with these fibres in this respect but are much stiffer than monolithic polymers, or in layman's terms plastics, which have low Young's modulus values of about 3 GPa.

How can we utilise the good strength and high stiffness of fibres? One possible solution would be to borrow an idea from natural organic fibre (e.g. hemp and şisal) technology and manufacture a ceramic/glass rope. However this solution has two major drawbacks.

(1) The strength of a rope is less than that of the fibres from which it is made (Fig. 20). There are a number of reasons for the disappointing performance of rope, the main being the fibres are wound helically and are therefore not pulled along their axes, the fibres only make up about 70% of the cross-section of the rope and the fibres rub against one another and introduce flaws.

STRENGTH
MPa

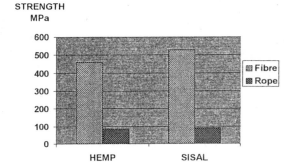

Fig. 20. Comparison of the strength of natural fibres and ropes.

(2) There is a severe limitation on the type of component you can make using rope.

How much better to separate the fibres from each other by embedding them in another material such as a polymer. In this way we can control the arrangement of the fibres and the manufacture of components with a wide range of shapes and dimensions is feasible. What we have made is a composite material consisting of two constituents — the fibres are known as the reinforcement and the polymer constituent as the matrix.

4. Composites

In a composite the load is taken by the fibres and the matrix, but mainly by the high stiffness and high strength fibres. In this way we can utilise the excellent properties of fibres and produce stiff, strong and tough material in a variety of forms. We have seen from the data of Table 2 that nowadays there are fibres with superior mechanical properties to glass which are commercially available for the production of composites. The best known of these are carbon fibres and these are used in a variety of high performance components, e.g. pole vaulter's pole and racing cars (Fig. 21). The improvement in mechanical properties achieved by reinforcing polymers is exemplified by the data for epoxy in Table 3. It is worthy of note that the superior mechanical

(a)

(b)

Fig. 21. Applications of carbon fibre reinforced polymers (a) McLaren racing car, the first carbon fibre reinforced plastic chassied Formula 1 racing car (b) pole vaulter's pole (courtesy Fiberite Europe).

Table 3. Comparison of the unidirectional mechanical properties of epoxies reinforced with various inorganic, synthetic fibres (Source: Dow Chemical Company).

	Young's Modulus (GPa)	Tensile Strength (MPa)
Epoxy	4	50
Epoxy + Glass	50	1165
Epoxy + Carbon	145	1480

properties of carbon fibres to those of glass fibres (Table 2) is reflected in the respective performance of their composites but even the latter have produced over an order of magnitude increase in both stiffness and strength. The worldwide production capacity of carbon fibres is huge, being estimated in 1991 to be over 13.5×10^6 kilogram per annum, but even this is small in comparison to glass fibre production. There is a considerable price differential between carbon and glass fibres so, in spite of their inferior properties, the less expensive glass fibres are still widely used. The importance of cost should never be forgotten; there is for example much current research on incorporating low performance, inexpensive natural fibres in polymers to give cheap mass produced components. As well as being a renewable resource, natural fibres are more readily recycled or disposed and already some car components such as inner trim parts (e.g. door panels, roof-liners and shelves) are made from a flax reinforced polymer.

The reinforcement in a composite need not be continuous fibres but can take many forms such as particulate and discontinuous fibres, as shown in Fig. 22. Not only may the form of the reinforcement differ but also the arrangement in the composite, for example discontinuous fibres may be aligned or randomly arranged.

The range of forms of reinforcement is exemplified by silicon carbide which is available as continuous fibres, whiskers (small crystals typically 20 μm in length and less than 1 μm diameter), platelets (Fig. 23) and equiaxed particles. Silicon carbide whiskers and platelets are produced commercially from a most novel raw material (Fig. 24). Yes — Fig. 24 shows rice growing in a paddy field. The husks of rice contain the raw ingredients for silicon

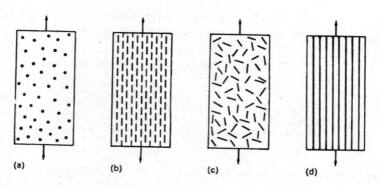

Fig. 22. Examples of composites (a) particulate (b) aligned discontinuous (c) random discontinuous (d) aligned continuous.[4]

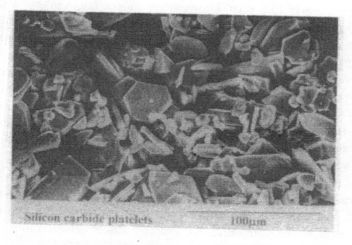

Fig. 23. Silicon carbide platelets (courtesy A. Selçuk).

carbide, namely the elements silicon and carbon, therefore all one has to do is to heat the rice husks in a carefully controlled manner to produce the whiskers and platelets. Typically the yield is 10% whiskers and 90% platelets. An organically grown inorganic reinforcement!

carbon from organic matter	+	mineral silica	=	silicon carbide	+	carbon monoxide gas
$3C$	+	SiO_2	=	SiC	+	$2CO$

Fig. 24. Silicon carbide whiskers and platelets are produced commercially from a most novel raw material — rice husks (courtesy Readers Digest).

(a)

(b)

Fig. 25. Reconstruction of the wall of a iron age hut; the wall is a hybrid composite.

5. Ceramic Matrix Composites

The main advances in composite materials have taken place over the last three decades but the concept of a composite material is not new and the walls of iron age mud huts were a composite material — a mud matrix reinforced by straw (Fig. 25). In fact, this was a very sophisticated composite because there are two forms of reinforcement — as well as the straw we also have branches as reinforcement. Composites with two types of reinforcement are called hybrid composites. A familiar example of a modern hybrid composite is the material used for car tyres, namely rubber reinforced with particulate carbon and continuous steel wires.

When mud dries it is hard and brittle and can be classed as a ceramic therefore the walls of a bronze age hut were made from a ceramic matrix composite (CMC). Ordinary concrete may be classified as a CMC with the cement as the matrix and the aggregate as the reinforcement. Some specialist cements are also fibre reinforced — asbestos and alkali-resistant glass fibres being commonly employed. However if we ignore these very high volume, low performance composites and just consider small volume, high quality composites then polymer matrix composites (PMCs) dominate the market. Nevertheless, although not so widespread as PMCs, there are a limited number of very successful "high-tech" ceramic matrix composites commercially produced. In these materials the main role of the reinforcement, which may be another brittle ceramic or a ductile metal, is to hinder crack propagation and so increase the toughness. There are numerous mechanisms by which a reinforcement may hinder crack propagation and a number of these may be operating and contributing to the toughness of a given CMC. Some of the more common mechanisms are illustrated in Fig. 26. Fibres are more effective in toughening a ceramic matrix than particulates as can be seen from the force-displacement curves, which are similar to stress–strain curves, of Fig. 27. Particulate reinforced CMCs fail in a catastrophic manner but they are tougher than their monolithic ceramic matrix counterparts as shown by comparison of the values for their work of fracture (area under the curve). Fibre reinforced CMCs retain some load carrying capacity after failure has commenced at the point of maximum force (stress) and this

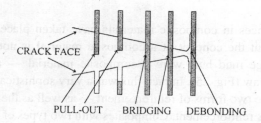

Fig. 26. Some toughening mechanisms that contribute to the toughness of composites.

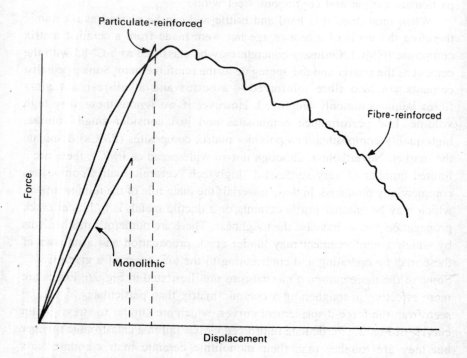

Fig. 27. Comparison of force-displacement curves from ceramic matrix composites and a monolithic ceramic.

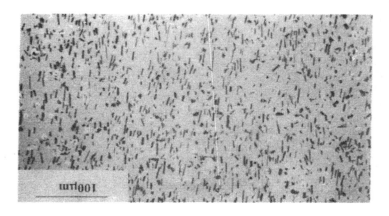

Fig. 28. Microstructure of zirconia reinforced with platelets of silicon carbide.[19,20]

is reflected in a large area under the force–displacement (stress–strain) curve.

Because of their superior properties to those of monolithic ceramics, ceramic matrix composites are used for components operating in severe conditions of stress, temperature and environment. I would like to discuss briefly two examples of ceramic matrix composites in commercial production.

Silicon carbide, in the form of equiaxed particles, platelets and whiskers, have been used to reinforced alumina and zirconia (Fig. 28). Zirconia based composites are not in full commercial production whereas SiC-reinforced alumina is used for cutting tools for wood and metal and give considerable productivity benefits over conventional tools. The SiC improves strength, toughness and thermal shock resistance without any adverse affects on other relevant performance indicators. Cutting tools are of course small components, about 10 mm × 10 mm × 5 mm, the next ceramic matrix composite is used for much larger components in some cases over 1 m in length (Fig. 29).

This is a more complex composite with a number of phases in its microstructure (Fig. 30). The main phases are alumina, which makes up most of the matrix, and graphite (carbon) flakes which are the reinforcement.[21]

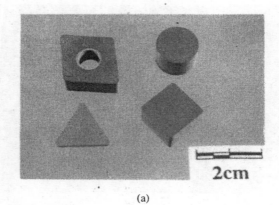

(a)

(b)

Fig. 29. Two examples of commercially available ceramic matrix composite components (a) SiC-reinforced alumina cutting tools (b) graphite-reinforced alumina components used in the steel industry.

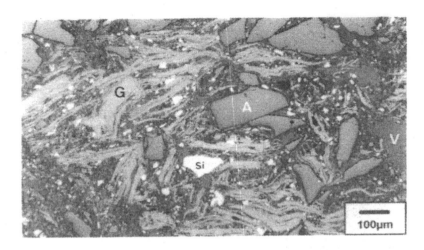

Fig. 30. Microstructure of the graphite reinforced alumina used for components such as the nozzles shown in Fig. 29; A is a alumina; Si, silicon; G, natural flake graphite; V, void (courtesy J.L. Leatherland, P.S. Rogers and R.D. Rawlings).

This graphite reinforced alumina is used in the steel industry, particularly in continuous casting plants where it has to cope with being in contact with corrosive, moving molten steel at about 1600°C. Its erosion, corrosion and thermal shock resistance are excellent, which is just as well as failure of a graphite reinforced alumina component would cause shut down of a continuous casting facility at a cost of £5,000 per minute!

The commercial ceramic matrix composites that we have considered so far had brittle reinforcements but it is also possible to toughen a ceramic by incorporating ductile metal fibres or particles. This form of toughening is never likely to be widely employed as the metal degrades some of the characteristic, useful properties of ceramics, for example it decreases hardness and reduces high temperature capabilities. However there are certain applications at room and intermediate temperatures that do not utilise the high hardness of ceramics for which a metal reinforced ceramic may be appropriate. Bioactive ceramics for medical implants would be such an application as the service temperature is constant at 37°C. The bioceramics

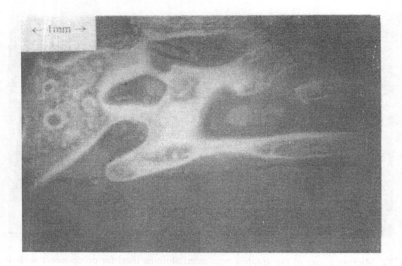

Fig. 31. Bone — glass-ceramic (Apoceram) interface showing good bonding (courtesy L.A. Wulff).

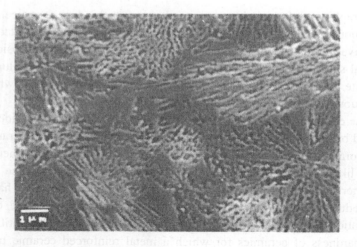

Fig. 32. Microstructure of Apoceram bioactive glass-ceramic (the crystalline phase is light coloured and the residual glass is dark).

discussed earlier that are used for the femoral head are classified as bioinert, which means that they are effectively ignored by the body. In contrast the body has an affinity for bioactive materials and consequently a strong chemical bond develops between a bioactive material and body tissue (Fig. 31). As a consequence of the bonding an implant made from a bioactive material does not have to be mechanically fixed (screwed or clamped) or cemented in position in the body. The bioactive material, Apoceram, shown in the Fig. 31 is a glass-ceramic which, as the name suggests, is a class of material that is related to both glasses and ceramics. The first stages in the production of a glass-ceramic is to produce a glass and to form the glass by casting or moulding into the required shape. The shaped article is then heat treated which causes the glass to transform in a controlled manner to the crystalline state.[22] The material is now a glass-ceramic and has a fine, polycrystalline microstructure (Fig. 32) and mechanical properties intermediate between a glass and a technical ceramic.

Bioactive glasses, glass-ceramics and ceramics share the undesirable characteristic with their conventional counterparts of being brittle. In order to improve the toughness we have reinforced bioactive glass-ceramics with particles of either titanium or silver, both these metals being acceptable to the body.[23,24] When a crack encounters a metal particle the particle deforms, which requires energy, and thus crack propagation is hindered and toughness is enhanced. Not only does the metal reinforcement improve toughness but slow crack growth, which can occur over a period of time when a component is under stress, is retarded.

In this last example we reinforced a brittle, hard ceramic with metal particles in order to improve the toughness. Can we change the roles of the two classes of materials and enhance certain properties of a metal, say hardness, by reinforcing with a ceramic?

6. Metal Matrix Composites and Functionally Graded Materials

So far we have discussed polymer and ceramic matrix composites, I would now like to consider metal matrix composites (MMCs). Typically reinforcing a metal with ceramic particles or fibres has the beneficial effects of increasing

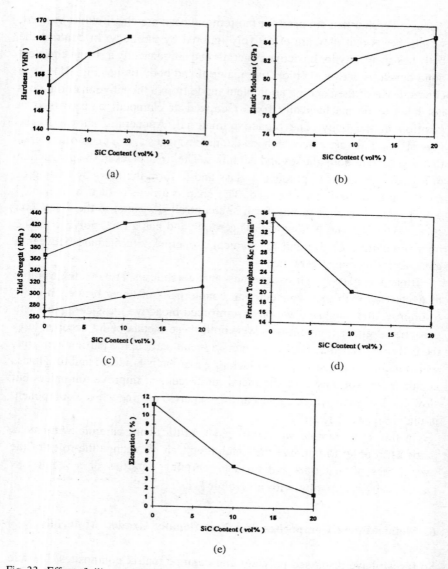

Fig. 33. Effect of silicon carbide content on the properties od Al2124 alloy matrix composites (a) hardness (b) Young's modulus (c) yield strength — tensile test •, bend test □ (d) fracture toughness and (e) ductility (courtesy C.-Y. Lin, H.B. McShane and R.D. Rawlings).

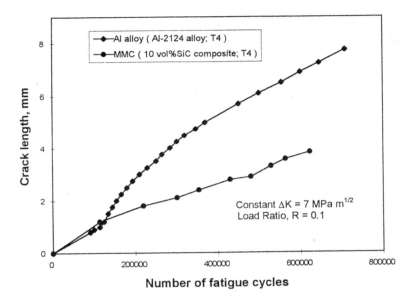

Fig. 34. Graph of change in crack length against number of cycles of the fluctuating stress demonstrating the slower rate of crack growth, i.e. the better fatigue crack resistance, of the MMC (courtesy H. Uzun, T.C. Lindley, H.B. McShane and R.D. Rawlings).

the hardness, stiffness, yield strength [Figs. 33(a), (b) and (c)], giving better fatigue resistance, that is resistance to crack growth under a fluctuating stress (Fig. 34), and improving high temperature performance. There is one major disadvantage with reinforcing a metal — you will recall that most metals are tough materials but unfortunately they become embrittled when we reinforce them; this is exemplified by the data for SiC particle reinforced aluminium alloy in Fig. 33(d) and (e). There is no doubt that the loss in toughness and ductility (strain to failure) is less marked when the processing of an MMC is carefully controlled so that the reinforcement is distributed evenly and not agglomerated and porosity and contamination are minimised; nevertheless, even with good processing the toughness of MMCs is inferior to their metal counterparts.

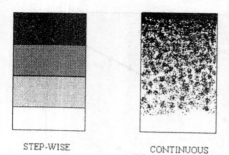

STEP-WISE   CONTINUOUS

Fig. 35. Schematic representation of functionally graded materials (a) stepwise (b) continuous.

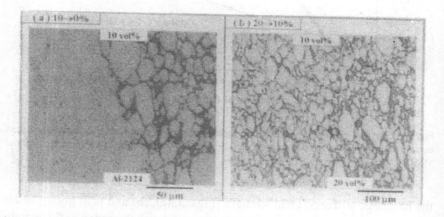

Fig. 36. Microstructures of SiC-Al alloy functionally graded materials (a) alloy and 10% SiC layers (b) 10% SiC and 20% SiC layers.[28]

How can we alleviate this problem? The approach we have taken is to produce materials with a gradation in reinforcement content; such materials are called functionally graded materials (FGMs). The gradation in reinforcement content may be stepwise or continuous (Fig. 35). The microstructures from a SiC-Al alloy stepwise FGM are given in Fig. 36; it can be seen that even with stepwise FGMs the interfaces between the layers are not well defined. By varying the reinforcement content in a specified manner we can

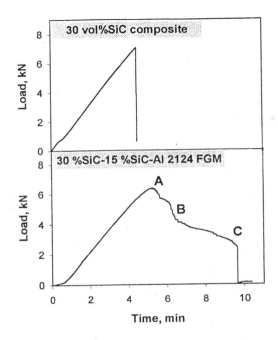

Fig. 37. Load-time curves (which are equivalent to load-displacement curves) from SENB tests on (a) SiC-Aluminium alloy composite and (b) FGM based on the SiC-Aluminium alloy system.[25]

obtain a predetermined variation in properties through the section of a component. For example, a component where the reinforcement content was high at the surface and decreased, in a stepwise or continuous manner, to zero towards the centre might benefit from the hard wear resistant surface, the good fatigue resistance of the reinforced near surface regions and the toughness of the unreinforced metal in the interior. Such an FGM would exhibit superior toughness to a composite material with the same reinforcement content as the surface region of the FGM. This is demonstrated by the load-displacement curves of Fig. 37; the load for the FGM does not fall catastrophically from the maximum, thus there is a larger area under the curve for the FGM.

7. Listening to Cracks

A material is tough when it is difficult for cracks to grow — it is brittle when crack propagation is easy. Clearly it would be useful to have a technique that is capable of detecting a crack and, if possible, detecting that crack during propagation. There are numerous techniques, which are classified as nondestructive testing techniques, that are designed to detect, and usually size, stationary cracks without impairing serviceability of the component. However there is only one readily available technique that can detect a growing crack but not an inactive crack; this unusual technique is called acoustic emission.

When we throw a stone into a pond it produces waves which propagate from the point of entry into the water to the edge of the pond. A similar phenomenon occurs when a crack propagates: elastic stress waves are produced at a growing crack and propagate through the material to the surface. In fact the reader will have heard elastic stress waves or acoustic emission, e.g. the noise of the cracking of a toughened glass windscreen or of ice. Another well known example of acoustic emission is the noise produced when zinc is deformed. In this case the acoustic emission is not due to cracking but to a mechanism of plastic deformation known as twinning, but the principle is the same. That zinc emits elastic stress waves has been known for centuries and has been used as an early nondestructive quality control test as illustrated by this excerpt from De La Pirotechnia, printed in Italy in 1540:

"That metal is known to be purer that shows its whiteness more, or if when it is broken it shows itself granular like steel inside, or if when some thin part of it is bent or squeezed by the teeth it gives its natural noise, like that which water makes when it is frozen by cold."

The acoustic emission from ice, toughened glass and zinc is of low frequency and high amplitude and therefore audible. In most cases the emissions are out of the range of the human ear and we have to use sensitive equipment to detect them. When an elastic stress waves reaches the surfaces there is a small, temporary surface displacement. This surface displacement can be detected and quantified by means of a sensitive transducer coupled

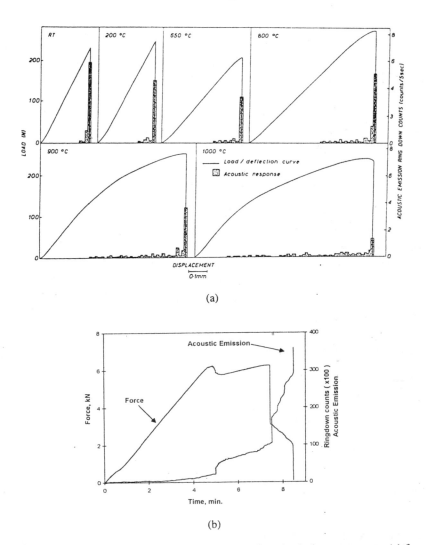

(a)

(b)

Fig. 38. Use of acoustic emission for the monitoring of mechanical property tests. (a) force (load) and acoustic emission rate as a function of displacement from tests on an impure alumina[27] (b) force and acoustic emission as a function of time (displacement) from a test on a stepwise FGM consisting of three layers, Al2124-30%SiC/Al2124-15% SiC/Al2124 (courtesy of H. Uzun, T.C. Lindley, H.B. McShane and R.D. Rawlings).

to the surface.[26] It is important to appreciate that the transducer does not have to be located close to the active crack as the elastic stress waves may travel considerable distances (in the case of metals several metres) and still produce detectable surface displacements. This may be easily verified by connecting a transducer to a long length of metal and introducing elastic stress waves at the other end by dropping a couple of grains of sand. Standard acoustic emission equipment will have no difficulty in detecting the artificially introduced stress waves.

Acoustic emission monitoring has been employed on all classes of materials, including ceramics and FGMs, examples of which are given in Fig. 38. Figure 38(a) shows that acoustic emission is sensitive enough to detect the small amount of crack growth, termed subcritical crack growth, that occurs before catastrophic crack propagation and failure in a ceramic at room and intermediate temperatures. The ceramic in this case was an impure alumina and at elevated temperatures ($\geq 650°C$) flow of a glassy phase associated with the impurities leads to nonlinearity in the force–displacement curves and a considerable amount of acoustic emission prior to fracture. Acoustic emission and load as a function of time (equivalent to displacement) from an SENB test on a stepwise FGM is presented in Fig. 38(b); it can be seen that both curves reflect the progress of the crack through the different layers.

The 16th Century Italian work on acoustic emission was mentioned earlier. I am pleased to report that I have evidence that the British were also aware of this phenomenon:

"The deed is done,
Did'st thou not hear a noise?"

This quotation is from Shakespeare's Macbeth!

Acknowledgements

I am greatly indebted to my wife, Ann, for her encouragement and support through every stage of my career. I would also like to thank the members

"Stress, depression and overwork, but enough about me - would you like to discuss your results in the two minutes I have before the staff meeting?"

of my research group for their efforts over the years. They not only have they carried out most of the work but have also been friendly and understanding; the following figure is dedicated to them (courtesy of The Times Magazine from which it was adapted).

References

1. An Introduction to Materials, Unit 6, *Solids Under Stress*, Open University Press, 1973.

2. C.F. Elam, The Distortion of Metal Crystals, Oxford University Press, 1935.

3. H. Carpenter and J.M. Robertson, Metals, Oxford University Press, 1939.

4. J.C. Anderson, K.D. Leaver, R.D. Rawlings and J.M. Alexander, Materials Science 4th Edition, Chapman and Hall, London, 1990.

5.   A.I.C. Persiano and R.D. Rawlings, Effects on Some Physical Properties of Soft Magnetic Fe-Co Alloys due to the Replacement of Vanadium by Niobium, *J. Mat. Eng.* **12**, 21, 1990.

6.   S. Hall, A. Mansur, H.-D. Pfannes, A.I.C. Persiano and R.D. Rawlings Mössbauer Spectroscopy of FeCo Alloys with Additions of Nb, Mo and Ta, *Proc. 1st Latin-American Conf. Applications Mössbauer Effect*, Rio de Janeiro, Brazil, eds. E. Baggio-Saitovitch, E. Galvão da Silva and H.R. Rechenberg, World Scientific, 263–266, 1988.

7.   A.I.C. Persiano and R.D. Rawlings, A Mössbauer Investigation of Equiatomic FeCo with Vanadium and Niobium Additions, *Phys. Stat. Sol.(a)*, **103**, 547, 1987.

8.   J. Dorner, Fashion, Octopus Books Ltd., London, 1974.

9.   R.D. Rawlings, The Manufacture of Cut-Steel Studs and Beads, *J. Historical Metallurgy Soc.* **12**, 88, 1978.

10. T. Gill, Gill's Technological and Microscopic Repository, Vol. 6, London, 1830.

11. B.J. Dalgleish and R.D. Rawlings, A Comparison of the Mechanical Properties of Alumina in Air and Simulated Body Environments, *J. Biomedical Mat. Res.* **15**, 527, 1981.

12. K.E. Aeberli and R.D. Rawlings, Effect of Simulated Body Environments on Crack Propagation in Alumina, *J. Mat. Sci. Letts.* **2**, 215, 1983.

13. M.W. Real, D.R. Cooper, R. Morrell, R.D. Rawlings, B. Weightman and R.W. Davidge, Mechanical Assessment of Biograde Alumina, *J. de Physique*, **47**, C1.763, 1986.

14. I. Thompson and R.D. Rawlings, Mechanical Behaviour of Zirconia and Zirconia-Toughened Alumina in a Simulated Body Environment, *Biomaterials*, **11**, 505, 1990.

15. R.D. Rawlings, Bioactive Glasses and Glass-Ceramics, *Clin. Mat.* **14**, 155, 1993.

16. G. Willmann, Zirconia — A Medical-Grade Material, *Proc. 6th. Int. Symp. Ceramics Medicine*, Philadelphia, USA, eds. P. Ducheyne and D. Christianson, *Bioceramics*, **6**, 271, 1993.

17. C.V. Boys, Production, Properties, and Uses of the Finest Threads, *Phil. Mag.* **23**, 486, 1887.

18. F.L. Matthews and R.D. Rawlings, Composite Materials: Engineering and Science, Chapman and Hall, London, 1994.

19. A. Selçuk, C. Leach and R.D. Rawlings, Processing, Microstructure and Mechanical Properties of SiC Platelet-Reinforced 3Y-TZP Composites, *J. European Ceram. Soc.* **15**, 33, 1995.

20. A. Selçuk, U. Klein, C. Leach and R.D. Rawlings, Processing and Microstructure of Pressureless-Sintered SiC Platelet-Reinforced Ce-TZP Composites, *Proc. Japan-Europe Symp. Composite Materials*, Nagoya, Japan, Japan Industrial Technology Association and R&D Institute of Metals and Future Industries, 257–262, 1993.

21. J.L. Leatherland, R.D. Rawlings and P.S. Rogers, Thermal Shock Testing of Alumina-Graphite Refractories, *Proc. 79th Steelmaking Conf.*, Pittsburgh, USA, Iron and Steel Soc. **79**, 409, 1996.

22. R.D. Rawlings, General Principles of Glass-Ceramic Production, Glass-Ceramic Materials-Fundamentals and Applications, eds. T. Manfredini, G.C. Pellacani and J.M. Rincon, *Series of Monographs on Materials Science, Engineering and Technology*, Mucchi Editore, Modena, Italy, 77, 1997.

23. E. Claxton, R.D. Rawlings and P.S. Rogers, The Mechanical Properties of a Bioactive Glass-Ceramic (Apoceram)-Titanium Composite, *Proc. 7th European Conf. Composite Materials* (ECCM-7 — Realising Their Commercial Potential), London, 1996, Woodhead Publishing Ltd., Cambridge, **2**, 443–448, 1996.

24. B.A. Taylor, R.D. Rawlings and P.S. Rogers, Development of a Bioactive Glass-Ceramic with the Incorporation of Titanium Particles, *Proc. 7th Int. Symp. Ceramics Medicine*, Turku, Findland, *Bioceramics*, eds. O.H. Andersson and A. Yli-Urpo, **7**, 255–260, 1994.

25. H. Uzun, H.B. McShane and R.D. Rawlings, Fabrication of SiC_p/Al-2124 Functionally Graded Materials by Hot Extrusion and Evaluation of Fracture Behaviour, *Proc. Ceramic Congress*, eds. V. Günay, J.H. Mandal and S. Özgen, *Turkish Ceramic Soc.* **2**, 186, 1997.

26. R.D. Rawlings, Acoustic Emission Methods, *Modern Techniques in Electrochemistry, Corrosion and Finishing*, eds. A.T. Kuhn, Wiley, 351, 1987.

27. B.J. Dalgleish, A. Fakhr, P.L. Pratt and R.D. Rawlings, The Temperature Dependence of the Fracture Toughness and Acoustic Emission of Polycrystalline Alumina, *J. Mat. Sci.* **14**, 2605, 1979.

28. C.-Y. Lin, H.B. McShane and R.D. Rawlings, Structure and Properties of Functionally Graded Aluminium Alloy 2124/Sic Composites, *Mat. Sci. Tech.* 659, 1994.

Professor Larry L. Hench

Dr Larry Hench was born on 21 November 1938 in Shelby, Ohio and graduated from The Ohio State University in 1961 and 1964 with BS and PhD degrees in Ceramic Engineering. He went to the University of Florida in 1964 as an Assistant Professor and in 1969 discovered Bioglass®, the first man-made material to bond with living tissues. In recognition of this discovery, which founded the field of bioactive medical and dental implants, and the development of methodology for investigating the interfacial bonding of this new class of biomaterials he was awarded the Clemson Award for Basic Research in 1977, the highest award of the Society for Biomaterials. This work combined with fundamental studies of glass systems led to a general theory of glass-environment interactions and the 1980 George W. Morey award, the highest honour of the Glass Division of the American Ceramic Society. The technology for manufacturing and quality assurance of bioactive glasses was achieved in Dr Hench's laboratory and transferred successfully to industry in 1984. Bioglass® prostheses for middle ear reconstruction, dental implants, and Bioglass® powders for repair for periodontal defects were approved for sale by the FDA in 1985, 1990, and 1993, respectively and are being marketed by US Biomaterials Corporation under license from the University of Florida. CE marks for sales of Bioglass® dental products in Europe were obtained in 1996 and in 1997 for orthopaedic products.

Dr Hench was promoted to Full Professor at the University of Florida in 1972 and to Graduate Research Professor of Materials Science and Engineering in 1986, the highest academic position at the University. He also served as Director for the Bioglass® Research Center and Co-Director of the Advanced Materials Research Center at the University of Florida for 15 years.

During 1978–1985, Professor Hench and his students conducted a series of basic science studies on glasses to immobilise high level radioactive wastes, including the first deep geological burial in Sweden. During this period he chaired key DOE and international committees leading to many government policy decisions in radioactive waste disposal.

In 1980, Professor Hench launched a new field of study — the low temperature sol-gel processing of glasses, ceramics, and composites. These studies have led to a new class of silica materials for optics and environmental sensors termed Gelsil®, which are being commercialised by Geltech Inc., a company founded in 1986 by Professor Hench and his wife and professional colleague, Dr. June Wilson Hench. The new gel-silica optics have received national recognition in the USA with 4 industrial awards including the prestigious 1991 R&D 100 award in optics.

Professor Hench's studies have resulted in 520 scientific publications, 23 books, and 23 patents issued in the US and 20 foreign countries. He is a member of the Academy of Ceramics and a Fellow in four professional societies. He has served numerous national offices including President of the Society for Biomaterials, Chairman of the Glass Division of the American Ceramic Society, Chairman of the Gordon Research Conference on Biomaterials, and is on the editorial board of 4 journals. He has chaired or cochaired 12 international conferences. His other awards include the 1982 Ceramic Education Council Outstanding Educator Award; Teacher-Scholar of the Year, the highest faculty award of the University of Florida; the 1983–1985 Florida Alumni Professorship, the 1985 State of Florida Scientist of the Year; the Samuel Scholes Award of the College of Ceramics at Alfred University, Outstanding Alumnus of the College of Engineering of the Ohio State University; the Wedgwood lecturer, the first non-British citizen so ho-noured, the Clemson University Hunter Lecturer and the Penn State Nelson Taylor Lecturer. He has received an Honorary Doctorate from Rose Hulman Institute of Technology. In 1998, he received the Materials Research Society Von Hippel Award, the highest award in materials science. He has supervised 50 PhD and 60 MSc graduates in Material Science and Engineering at the University of Florida. He is married and he and his wife have two sons, two daughters and nine grandchildren. His hobbies include oil painting, photography, snorkelling, and writing children's storybooks which are being published by The American Ceramic Society.

In 1996, Dr Hench accepted a University of London Chair at Imperial College of Science, Technology and Medicine as Professor of Ceramic Materials. He also serves as Director of the Imperial College Centre for Tissue Regeneration and Repair. At Imperial College he is actively involved in creating a multidisciplinary approach to teaching and research in tissue engineering, artificial organs and materials for regeneration of diseased, damaged or ageing tissues.

29 October 1998

THE STORY OF BIOGLASS: FROM CONCEPT TO CLINIC

LARRY L. HENCH

Department of Materials
Imperial College of Science
Technology and Medicine
Prince Consort Road
London, SW7 2BP, UK
E-mail: l.hench@ic.ac.uk

1. The Beginning

The story of Bioglass® perhaps began many eons ago when life first emerged from the seas. The earth cooled to form the geosphere 5 to 6 billion years ago. Silica sand and water eventually became the two most abundant chemical compounds on the earth's surface. By mechanisms that are still more myth than science, life was created and order emerged from chaos.[1] As Schrödinger said many years ago; "Life — the great mystery, it sucks order from disorder." The order, we now know, is stored in DNA and RNA. All species of life, from the simplest single cell bacteria to vertebrates, such as humans, possess the same organic constituents which appear to be traceable back to a single origin.

The earth's biological clock, based upon the fossil record, shows that following some 2 billion years of trial and error a momentous change in biology took place and is retained in hundreds of thousands of species today.

®Registed Trademark University of Florida, Gainesville, Florida.

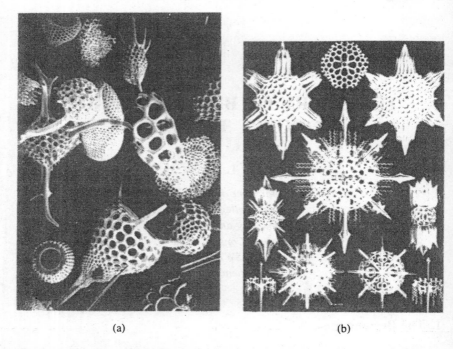

(a)	(b)

Fig. 1. Marine organisms with hydrated silica exoskeletons. (a) Diatoms (b) Radiolarians.

This dramatic event was the biochemical process of building exoskeletons from inorganic elements. Diatoms and radiolarians are just two examples of life forms which learned to protect themselves by building intricate silica-based skeletons (Fig. 1).

A diatom builds its frustule, or skeleton, using a relatively simple protein template or pattern. The building blocks are molecules of hydrated silica found in parts per million in all oceans and seas.

A recently published model of the biomineralization of a diatom frustule by K. Lobel, J. West and myself[2] shows the mechanism by which a protein template can epitaxially nucleate a stable 3-D structure of silica. Hecky, a marine biochemist, and colleagues showed nearly 20 years ago that the organic layer of a diatom frustule contains a large concentration of the amino acid serine.[3] Molecular mechanics shows that the unpaired bonds of a polyserine Beta sheet exactly match the terminal silanols of a 4-membered tetrasiloxane ring.

Semiempirical molecular orbital (MO) calculations show that once 4-membered silica rings are formed they are thermodynamically stable[4] (Fig. 2). Thus, if the correct template is present, which could have been formed by evolutionary chance, only a simple low energy inorganic condensation reaction is required to nucleate and grow an intricate silica structure. Another set of MO calculations by J. West and myself has shown that a hydrated silica cluster, such as is present on the surface of many silica-containing minerals, can serve as an inorganic catalytic substrate for synthesis of peptide bonds[4] (Fig. 3). The polypeptides can then serve as a template for further self-assembly of organic molecules, one of the requirements for life.[1]

Evidence of the feasibility of forming silica structures at sea temperatures is found in the use of sol-gel chemistry by my group of researchers at the University of Florida to make pure silica optics.[5,6]

All of these optical components were made by hydrolysis-condensation reactions of silica precursors leading to gelation at ambient temperature (Fig. 4). Casting the inorganic sol in the presence of an organic template creates diffractive optic patterns on the silica with a scale of micrometers identical to that of diatoms and radiolarians (Fig. 5).

Biomineralisation of silica was an immensely important and effective step forward in evolution because 60% of all nitrogen is now fixed biochemically by such species. They begin the food chain which culminates in us.

We learn two major principles of biology from examining the beginnings of skeletal formation. First, form and function are intimately connected. Secondly, there is a vital equilibrium between the biomechanical and biochemical requirements of a lifeform. Inorganic and organic chemistry function concurrently to create materials which preserve this equilibrium.

Much later in the biological clock organisms learned how to process calcium salts as well as silica to form internal skeletons. Such species offered additional versatility and have prospered greatly in the biosphere. Very complicated structures of calcium phosphate salts, similar to the mineral

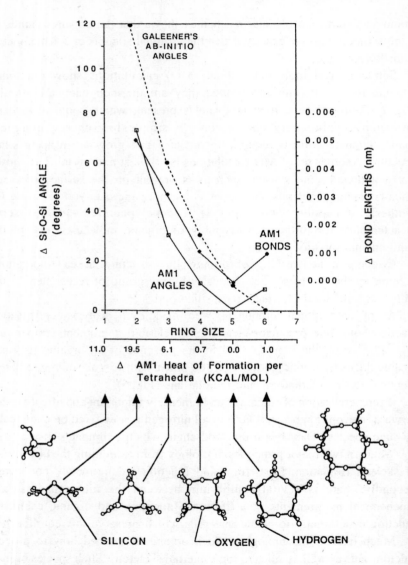

Fig. 2. The structure and thermodynamic stability of silica clusters with one to six tetrahedra per cluster calculated with AM-1 semi-empirical quantum mechanical models. Note that silica clusters with four, five or six tetrahedra have nearly equivalent heats of formation.

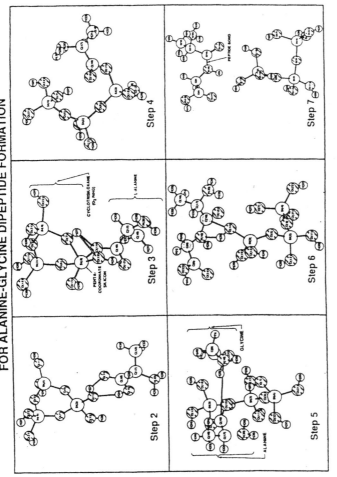

(a)

Fig. 3. Reaction pathway for a trisiloxane (3-member) silica ring acting as an inorganic enzyme for formation of an alanine-glycine peptide bond with low activation energy barrier. (a) Molecular configuration of critical reaction steps in bond formation. (b) Summary of the energetics of the 13 step reaction path of peptide bond formation.

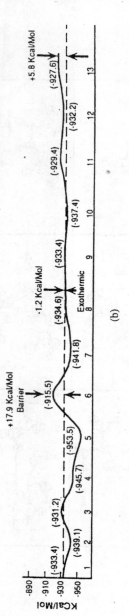

Fig. 3 (*Continued*)

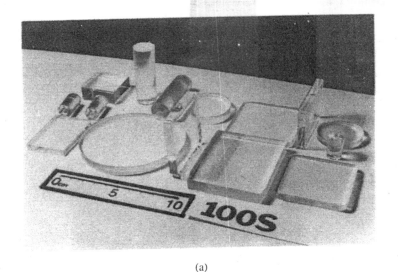

(a)

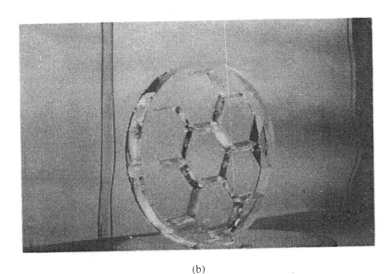

(b)

Fig. 4. Silica optical components made by low temperature sol-gel processing. (a) Transmissive optical elements made of 100% gel-silica. (b) Large 10 cm lightweight optical mirror made by sol-gel processing of silica.

(a)

(b)

(c)

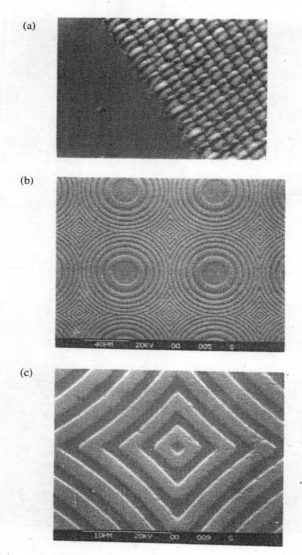

Fig. 5. Surface diffractive optics made by replication of polymer molds in gel-silica. (a) lenslet array with spacing of a few micrometers. (b) Diffractive optic array of approximately 40 micrometer scale. (c) Diffractive features replicated at the sub-micrometer scale. Note similarity of scale and structural features to the diatoms and radiolarians of Fig. 1.

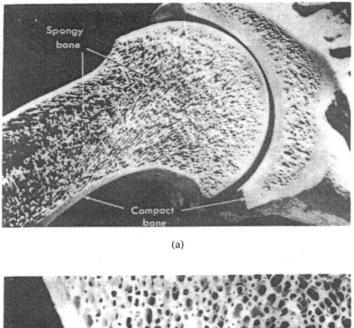

(a)

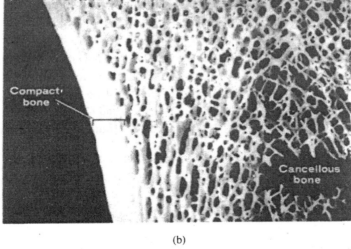

(b)

Fig. 6. Structure of cancellous (spongy or trabecular) bone with compact (cortical) bone. (a) In the hip joint. (b) Expanded scale view of the structure of cancellous bone. Note the similarity of architecture with the diatoms and radiolarians of Fig. 1 and the silica optics structures of Figs. 4 and 5.

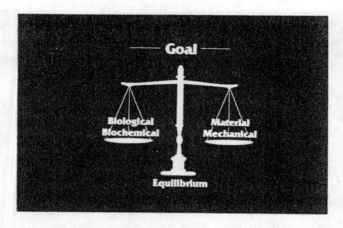

Fig. 7. The goal of biomaterials is to restore the equilibrium between biomechanical loads applied to tissues and the biochemical signals that maintain health of the tissues.

hydroxyapatite, were eventually created. Two of the most important, cortical bone and trabecular bone, provide a composite structure that is both strong and light weight (Fig. 6). The scale of the cells and the extracellular matrix associated with endoskeletal bone are similar to the scale of the structures that comprise the inorganic exoskeletal structures of diatoms and radiolarians. There is an exquisite balance between structural scale (form) and structural function in all biological systems. Biomechanics and biochemistry are linked together at the cellular level and are in equilibrium when an organism is healthy (Fig. 7).

However, the equilibrium of biomechanics with biochemistry and the balance between form and function of bone begins to deteriorate with age.

2. The Problem

Our skeletal structures reach maturity around twenty years of age and for another ten years equilibrium is maintained. However, from the age of 30 years onwards the continual repair and remodeling of bone to stresses and fatigue damage becomes progressively less efficient. By the age of 60, the cross sectional area of trabecular bone has decreased by 15–30% for men and as much as 25–40% for women, as illustrated in [Fig. 6(b)] for a section of trabecular bone from a 60 year old women. Note the decrease in thickness of the trabeculae and the missing segments. These changes correspond to a large decrease in bone mass which in turn leads to a reduction in strength and a large increase in the probability of fracture of the bone. Elderly bone, especially osteoporotic bone, is especially vulnerable to fracture of the long bones and the femoral neck (a broken hip) and collapse of vertebrae. Thus, restoration of skeletal equilibrium is necessary to maintain mobility and an acceptable quality of life. Implants made from special compositions of materials, called biomaterials, are now commonly used to repair or replace bones, joints and teeth. The use of biomaterials in the body has revolutionised health care for the aged and has provided a better quality of life for millions of people.

One of the great challenges facing materials science and technology today is the development of a new generation of biomaterials to repair the body. The problem is the need for materials which will survive as long as the patient using them. This is often 15–30 years, double the expected lifetime of many spare parts in use today.[7] A new approach is needed to control the chemistry of materials so that they will exhibit longer lifetimes in the body. Materials need to be designed for patients with debilitating skeletal diseases, such as osteoporosis and arthritis. Self-repair, a characteristic of living tissues, is not possible with the materials in use today. Eventually materials need to be developed that will aid in the regeneration of tissues rather than replace them.

For many years the guiding principle used in the use of biomaterials was that the material should be as chemically inert as possible.[8] Body fluids are highly corrosive saline solutions. The first materials used in skeletal repair were metals optimized for strength and corrosion resistance. Metallic implants for orthopedic applications have been very successful with hundreds of thousands being implanted annually.[7] The original applications were as removable devices, such as those for stabilisation of fractures. Use as permanent joint replacements began in the 1960s with the development of self-curing polymethylmethacrylate "bone cement" which provided a stable mechanical anchor for a metallic prosthesis in its bony bed.[7] High levels of clinical success (>75% over 25 years) of "cemented" metallic orthopedic implants, especially for total hip prostheses, have led to rapid growth in the use of implants.

The increase in number of implants has been accompanied by an increase in the life expectancy of patients and a decrease in the average age of patients receiving an implant. This means that a growing proportion of patients will outlive the expected lifetime of their prostheses.[7]

Many factors can contribute to the failure and lifetime of an implant; however, instability of the interface between the implant and its host tissue is one of the most critical problems.[8,9] The primary causes of interfacial instability are chemical and mechanical mismatch between the implant and living tissues. The body responds to nearly bioinert metallic implants by developing a thin, nonadherent fibrous capsule that isolates the device from its host tissue. The implant can move within the capsule and create local stress concentrations at the interface leading to loosening, wear and wear products, pain and even fracture of the bone or implant.

Metallic implants have elastic moduli which are many times higher than bone. The mismatch of elastic modulus means that most of the load of body weight is carried by the implant. This is bad because bone must be continuously loaded in tension in order to remain healthy. Under compressive loads, or no load, bone resorbs through a complex cellular process.[10] The effect of bone resorption is to decrease the amount of bone in contact with the prosthesis. This weakens the support structure for the device and increases the probability of failure even further. This problem is called stress shielding and it gets worse as we age because of a diminished ability to grow new bone in response to mechanical stress.

3. The Concept: Bioactive Bonding

In 1967, I was introduced to the problem of surgical rejection of metallic and polymeric implants by a medical doctor who had recently returned from a battlefield assignment in Vietnam. Colonel Klinker was in the United

States Medical Research and Development Command. During a bus ride to an Army Materials Research Conference in Sagamore, New York, he described the difficulty in repairing long bone injuries with the orthopaedic materials and devices then available. He recalled many cases that he had experienced of large sections of bone lost due to impact by high velocity bullets or land mines. He said, "What is needed is a material that the body will not reject. All metals and plastics are rejected by formation of scar tissue around them. Why don't you develop a ceramic material that the body won't reject?"

In 1969, the US Army Medical R&D Command funded a one year proposal, which I directed to explore the hypothesis that a glass or glass-ceramic material containing calcium and phosphate ions, the same constituents found in bone mineral, would not be rejected by the body. From this hypothesis evolved a new concept and the discovery as to how to provide a stable interface between living and nonliving materials.[11]

The new concept to achieve a stable implant-tissue interface is use of *bioactive fixation*. A bioactive material is one that elicits a specific biological response at the interface of the material which results in the formation of a bond between the tissues and the material.[12] This concept is based upon control of the surface chemistry of the material and its interface with tissues. A bioactive implant reacts chemically with body fluids in a manner that is compatible with the repair processes of the tissues. A fibrous capsule is prevented from forming by the adhesion of repairing tissues. Since the chemical reactions are restricted to the surface, the material does not degrade in strength as do resorbable or porous implants.

The first bioactive material reported was a simple four component glass composed of SiO_2, Na_2O, CaO, and P_2O_5 (composition 45S5 in Table 1).[11,12]

Table 1. (in weight percent).

SiO_2	Na_2O	CaO	P_2O_5
45	24.5	24.5	6

The low silica content and presence of calcium and phosphate ions in the glass result in very rapid ion exchange in physiological solutions and rapid nucleation and crystallization of hydroxyl carbonate apatite bone mineral on the surface. The growing bone mineral layer bonds to collagen, produced by the bone cells, forming a strong interfacial bond between the inorganic implant and the living tissues. During the last thirty years numerous materials have been shown to develop bioactive bonding to bone, but the amazing fact is that no material bonds more rapidly than the first compositions tested in November, 1969.[13]

The interfacial strength developed by most bioactive materials is equivalent to or greater than that of bone.[14] After implantation in bone and testing to failure, fracture occurs in bone or the implant material depending upon their relative strength. Failure does not occur through the interface. A few compositions of bioactive glasses (45S5 Bioglass® in Table 1) also bond to soft tissues, as well as bone, with an adherence strength greater than the cohesive strength of the collagen fiber bundles of the soft connective tissues, as discovered by June Wilson in 1980.[15,16]

4. Bioactive Glasses

Bioactive implants have differing rates of bonding depending upon their composition. The most rapid rates of bonding occur for bioactive glasses with SiO_2 contents of 45–52 weight %. Soft tissue as well as bone bonding takes place within 5–10 days for these compositions. Glasses with these high rates of bonding also enhance the rate of bone proliferation, called *osteoproduction*, also discovered by Dr June Wilson.[17] These materials are considered to exhibit Class A bioactivity.[13] Bioactive glasses or glass-ceramics containing 55–60 weight % SiO_2 require longer to form a bond with bone and do not bond to soft connective tissues. Synthetic hydroxyapatite ceramics also bond to bone but slowly; bone slowly grows along the implant interface. This behaviour is termed *osteoconduction* and such materials are considered to exhibit Class B bioactivity. The compositional range of bioactive glasses that exhibit Classes A and B bioactivity are shown in Fig. 8. Outside of the bioactive bone-bonding boundary the glasses are bioinert and elicit formation

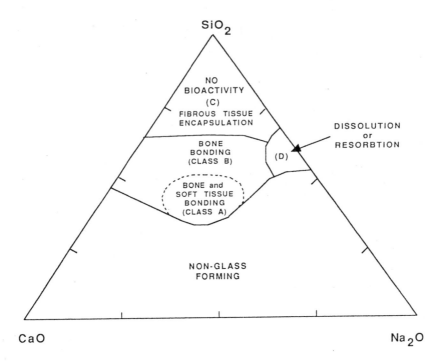

Fig. 8. Compositional diagram, in weight percent, of the region of Class A bioactivity where both bone and soft tissue bonding occurs, the region of Class B bioactivity where only bone bonding occurs, the region of no bioactivity and the region of resorption. Note: all compositions shown contain an additional 6% P_2O_5.

of a nonadherent fibrous capsule. Compositions without sufficient calcium ions are too reactive and resorb in the body.

The reason for these important differences in *in vivo* behaviour between various bioactive implants is related to their surface reaction kinetics in physiological solutions. Figure 9 summarises the sequence of twelve reactions which occur on the surface of a bioactive glass as a bond with bone is formed. The level of understanding of these processes is quite high for Stages 1–5 but is sparse for Stages 6–11. The limitation in several stages, such as 8 and 9, is the lack of basic knowledge of the biological processes that

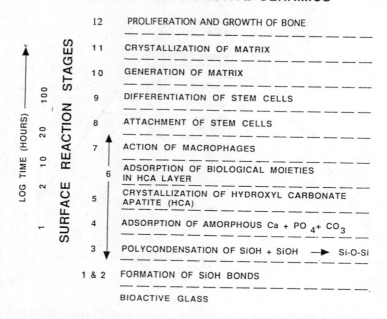

Fig. 9. Sequence of interfacial reactions and their time dependence in forming a bond between bone and a Class A bioactive glass.

control the genetic expression of highly differentiated cells like osteoblasts. Bioactive substrates with known differences in surface chemistry and surface kinetics are unique model systems for exploring these biological phenomena.

The time scale for the surface reactions for a glass with high bioactivity is also shown in Fig. 9. Glasses with the highest levels of bioactivity undergo surface reaction Stages 1–5 very rapidly. A polycrystalline hydroxyl carbonate apatite layer is formed on 45S5 Bioglass® (Stages 4 and 5), for example, within 3 hours both *in vitro* in simulated body fluids and *in vivo*. In contrast, compositions with intermediate, Class B, levels of bioactivity that bond only to bone require 2–4 weeks to form a crystalline hydroxyl carbonate apatite layer on the material. Compositions that are not bioactive and do not form a bond to either bone or soft tissues do not form a crystalline apatite layer even after 4 weeks in solution.

The differences in rates of change of the inorganic phases on the surface of the material alter which biological species are adsorbed (Stage 6 in Fig. 9) and subsequently the rate and types of interfacial cellular responses (Stages 7–11 in Fig. 9). For example, the attachment and spreading time of certain types of fibroblasts, which can form a nonadherent fibrous capsule at an implant interface, are slowed down considerably on a bioactive glass surface compared with inert surfaces whereas the cellular activity of bone-growing cells (osteoblasts) is enhanced.

Bioactive materials offer several advantages with respect to the development of a new generation of implants with enhanced lifetimes: (1) opportunity for molecular tailoring of compositions to match the biochemical requirements of diseased and damaged tissues, (2) design flexibility to tailor biomechanical characteristics to match those of the natural host tissues by optimising the microstructure of the material or use as a second phase in a composite.

Reference 4 summarises the physical properties and compositions of various bioactive ceramics and shows that the potential versatility of this new class of biomaterials is beginning to be realised. Bioactivity is retained even in multiphase glass-ceramics. Strength and toughness have been greatly increased since the first bioactive glasses were reported. Thus, the composition of a bioactive matrix phase can be optimised for chemical reaction rates to match physiological requirements with an inactive reinforcing phase optimised with respect to size and distribution, thereby maximising strength and toughness. A/W glass-ceramic, composed of apatite and wollastonite crystals in a silicate glass matrix, developed by Yamamuro and Kokubo, has greatly enhanced mechanical properties over 45S5 Bioglass®, for example.[18,19]

5. Four Paths of Development

During the 30 years since the concept of bioactivity was discovered there have been four paths of development followed, as summarised in Fig. 10. *Path A* has emphasised understanding of the Mechanisms of Bioactive

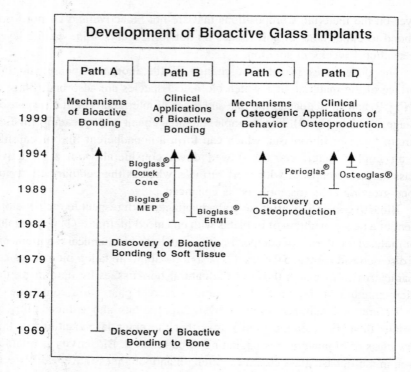

Fig. 10. Four pathways of development of bioactive glass implants from their discovery in 1969 by L.L. Hench and colleagues.

Bonding. Many new analytical techniques were developed to investigate the chemical and ultrastructural nature of the bonded interface. The 12 reaction stages summarised above are based upon this extensive series of investigations of the surface kinetics of bioactive glasses and the effect of environmental variables on the rates of reaction.

Path B is the Clinical Applications of Bioactive Glasses. During the 12 years since clinical use was launched excellent clinical success is reported for the bioactive glasses in nonloaded applications such as middle ear implants and alveolar ridge maintenance implants.[20-23]

There have been no extrusions of the middle ear prostheses through the ear drum, the most common mode of failure of metallic and plastic middle ear prostheses.[7] Also, the alveolar ridge dental implants made from Bioglass® seldom have been lost or have had to be removed due to resorption of the jaw bone around them, unlike the performance of Class B, synthetic hydroxyapatite implants used in the same manner. The advantage of soft and hard tissue bonding provides the equilibrium of stress transfer required to maintain the health of bone, as described earlier.

Bioactive A/W glass-ceramic implants are used very successfully in vertebral repair and replacement of portions of the iliac crest in total hip surgery.[19,24] As yet very little data exist on the environmental sensitivity and fatigue life of such multiphase materials under the complex physiological loads of total joint replacements and they are not used in these situations. The elastic moduli of all of the present generation of bioactive ceramics are too high to avoid the problem of stress shielding. Specially designed bioactive composites appear to be the best way to optimise biomechanical properties with a biochemically optimised surface.

6. Bioactive Composites

Many research teams are attempting to develop composites and coatings that achieve improved interfacial stability and mechanical properties. (See Yamamuro, Hench and Wilson, Vols. I and II for a large collection of papers and reviews of these materials.)[24,25] The interest in this approach to the interface instability problem is largely responsible for the large growth in research centres worldwide studying bioactive ceramics, which has reached a number of 85 laboratories in 15 countries.

Figure 11 summarises the properties of some of the more promising composite biomaterials and compares them with other biomaterials used clinically. Few of the systems studied eliminate the critical problem of stress shielding. Professor Bonfield's group has made progress in this area using a higher modulus bioactive phase of hydroxyapatite (HA) powders dispersed in a low modulus high density polyethylene matrix.[26,27] The resulting composite has a Young's modulus from 1 to 8 GPa and a strain

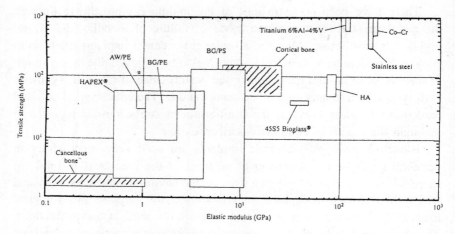

Fig. 11. Comparison of the mechanical properties of various biomaterials with cortical and cancellous bone. Hapex® is a composite of hydroxyapatite in a polyethylene matrix, AW/PE is a composite of bioactive apatite-wollastonite glass-ceramic particles in a polyethylene matrix, BG/PE is 45S5 Bioglass® particles in a polyethylene matrix, BG/PS is a bioactive composite of 45S5 Bioglass® particles in a polysulfone matrix, HA is synthetic hydroxyapatite. Note that only the BG/PS composite comes close to matching the mechanical properties of compact, cortical bone.

to failure from >90% to 3% as the volume fraction of HA increases to 0.5. The transition from ductile to brittle behavior occurs between 0.4 and 0.45 volume fraction of HA. The next step in development of an optimised two phase composite is to use an inorganic phase with higher rates of bioactivity. PE-bioactive glass composites are currently being developed in The Interdisciplinary Research Centre in Biomedical Materials at the University of London which may yield the required combination of chemical and mechanical characteristics for load bearing devices with enhanced lifetimes, interfacial bonding and minimal stress shielding.[28,29] Poylsulfone-Bioglass® composites have been jointly developed at the University of Florida and my new research group at Imperial College.[30] Properties of these new molecularly tailored bioactive composites are included in Fig. 11. Only time will tell whether clinical use of bioactive composites as load-bearing devices is practical. Fatigue data in a biological environment under physiological loads is still to be obtained.

7. Molecular Tailoring of Surface Chemistry

The greatest challenge for inorganic biomaterials is the design of the surface chemistry of materials to meet the requirements of aged, diseased or damaged tissues. Most of the materials in use today were developed by trial and error. There are almost no data on the *in vivo* response of the materials as a function of the age of tissue or the effects of disease states, such as osteoporosis or arthritis, on interfacial reactions or biomechanical behavior of interfaces. It is only in the last few years that some principles have been established to guide the development of new materials. Surface reaction kinetics of bioactive ceramics of various compositions have been determined (Stages 1–5 in Fig. 9). The compositional effects responsible for controlling the reaction rates are now known or reasonably well understood.[4] However, details of Stages 6–11 in Fig. 9 are only beginning to be investigated. The materials variables responsible for controlling the *in vivo* bioactivity of a material are poorly understood. This lack of understanding of the bone-materials interface is a major barrier to the design of new materials with improved performance and has recently become the focus of many research programs.

Two new directions of research hold promise for improving the scientific basis for tailoring surface reactions of inorganic biomaterials indicated by *Paths C* and *D* in Fig. 10. One is the discovery that sol-gel derived glasses have a much expanded compositional range of bioactivity over glasses made by traditional melting and casting processes.[31,32] Sol-gel processing is one of the most important new methods for production of new, chemically derived materials.[5,33] The low temperatures of sol-gel processing, make it possible to control surface chemistry of the resulting materials with greater flexibility than high temperature melting and casting of glasses or sintering or hot pressing of ceramics. Details of the seven processing steps in making bioactive gel-glasses are discussed in Refs. 33 and 34. Advantages of sol-gel processing of inorganic biomaterials include: new compositions, greater homogeneity,

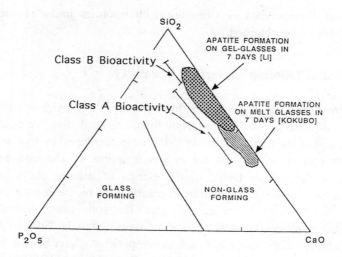

Fig. 12. Expanded compositional range of bioactive gel-glasses compared with melt glasses from the same system.

higher levels of purity, net-shape casting of monoliths, low temperature coating of substrates, control of powder size distribution, control of surface chemistry of the gel-glasses, expanded ranges of glass formation, control of pore networks at a nanometer scale, in addition to commercial advantages such as lower energy consumption and nearly zero environmental impact.

Figure 12 shows the extended range of compositions in the SiO_2-CaO-P_2O_5 system that are bioactive when made by alkoxide based sol-gel processing compared with bioactive compositions made by melting and casting. Gel-derived glasses with as much as 88% SiO_2 develop hydroxyl carbonate apatite (HCA) layers whereas the limit for melt-glasses is 60%. This is a huge shift in compositional boundary for bioactivity. Melt-glasses with >55% SiO_2 require several days to form a polycrystalline HCA layer whereas gel-glasses do so in only a few minutes.[4,31,32] The chemical origin of these important differences appears to be the large concentration of silanols on the surface of the gels after processing temperatures of 500–800°C. Semiempirical molecular orbital (MO) calculations, using AM-1 methods, show that metastable silica clusters formed from a condensation reaction of neighboring silanols (Stage 3 in Fig. 3) can act as heterogeneous nucleation sites for hydroxyl carbonate apatite crystals (Stages 4 and 5).[4] The metastable silica clusters can also act as preferential adsorption sites for amino acids, such as alanine.[4,35] These calculational results indicates that the surface reactions of the inorganic material (Stages 1–5) can lead to biologically specific binding sites for protein molecules (Stage 6). The MO calculations show differences in specific adsorption on the inorganic surface which depend on different binding sites on the protein molecules. This may lead to an understanding of the selective adsorption of proteins that act as growth factors or enzymes (Stages 6–8). Such studies may also aid in the interpretation and optimisation of new hybrid inorganic-biological systems, such as alkaline phosphates, enzymes or other optically active organic molecules trapped within sol-gel silica porous glass matrices.[4]

8. Conclusions

The story of Bioglass that we hypothesised began when living organisms discovered how to build chemically stable exoskeletons from soluble silica has now come full circle after several billion years.[36,37] We are using the biological behaviour of man-made glasses and that release controlled concentrations of soluble silica to influence the stability and enhance the regeneration of living tissues.

Results from molecular orbital modeling calculations, combined with experimental investigations of the adsorption of biological growth factors and other biological species, should make it possible to design a new generation of gel-glasses that enhance the rates of interfacial bonding and regeneration of even aged or diseased tissues. Bioactive composites that have high toughness and a modulus of elasticity that will prevent stress shielding of bone are also now feasible. The potential solutions to lifetime problems of prostheses lie in the creative use of materials chemistry. Millions of people will benefit if this potential can be realised. The research direction to follow is finally apparent after many years of trial and error. It took 1 to 2 billion years of trial and error for the diatoms to emerge triumphant in the seas. Hopefully we can speed up that process in improving the quality of life for ageing humans.

Acknowledgements

The author gratefully acknowledges the financial support during thirty years of the US Army Medical R&D Command and the US Air Force Office of Scientific Research Division of Chemical and Materials Sciences. The long standing assistance of Dr. June Wilson, Alice Holt, and Dr. Jon West is also gratefully acknowledged along with the research of dozens of creative graduate students.

References

1. R. Shapiro, *Origins: A Skeptics Guide to the Creation of Life on Earth.* Penguin Books, New York, 1988.

2. K. Lobel, J.K. West and L.L. Hench, A Computational Model for Protein-mediated Biomineralisation of the Diatom Frustule. *Marine Biology*, **126**, 353, 1996.

3. R.E. Hecky, *Marine Biology*, **1973**, 19, 1973.

4. L.L. Hench and J.K. West, Biological Applications of Bioactive Glasses. *Life Chemistry Reports*, **13**, 187, 1996.

5. L.L. Hench and J.L. Nogues, Sol-Gel Processing of Net Shape Silica Optics. *Sol-Gel Optics*, ed. Lisa C. Klein, Kluwer Academic Publishers, Norwell, MA, 59, 1994.

6. T. Chia and L.L. Hench, Laser Densification of Micro-Optical Arrays. *Sol-gel Optics: Processing and Applications*, ed. Lisa C. Klein, Kluwer Academic Publishers, Norwell MA, 511–538, 1994.

7. L.L. Hench and J. Wilson, *Clinical Performance of Skeletal Prostheses.* Chapman & Hall, London, 1996.

8. L.L. Hench and E.C. Ethridge, *Biomaterials: An Interfacial Approach.* Academic Press New York, New York, 1982.

9. L.L. Hench and J. Wilson (eds.), *Introduction to Bioceramics.* World Scientific Publishers, London and Singapore, 1993.

10. P. Revell, *Pathology of Bone*, Springer-Verlag, Berlin, 1986.

11. L.L. Hench, R.J. Splinter, T.K. Greenlee and W.C. Allen, Bonding Mechanisms of the Interface of Ceramic Prosthetic Materials. *J. Biomed. Mat. Res.* **No. 2**, Part 117–141, November 1971.

12. L.L. Hench and H.A. Paschall, Histo-Chemical Responses of a Biomaterials Interface. *J. Biomed. Mat. Res.* **No. 5**, Part 1, 49, 1974.

13. L.L. Hench, Bioactive Ceramics: Theory and Clinical Applications. *Bioceramics* 7, eds. O.H. Anderson and A. Yii-Urpo, Butterworth-Heinemann Ltd., Oxford, England., 3–14, 1994.

14. L.L. Hench, H.A. Paschall, W.C. Allen and G. Piotrowski, Interfacial Behaviour of Ceramic Implants. *National Bureau of Standards Special Publication*, **41**, 129, 15 May 1975.

15. J. Wilson, G.H. Pigott, F.J. Schoen and L.L. Hench, Toxicology and Biocompatibility of Bioglasses®. *J. Biomed. Mat. Res.* **15**, 805, 1981.

16. J. Wilson and D. Nolletti, *Handbook of Bioactive Ceramics*. Vol. I, eds. T. Yamamuro, L.L. Hench and J. Wilson, CRC Press, Boca Raton, Florida, 1, 283–302, 1990.

17. J. Wilson and S.B. Low, Bioactive Ceramics for Periodontal Treatment. Comparative Studies in the Patus Monkey. *J. Appl. Biomat.* 3, 123, 1992.

18. T. Yamamuro, A/W Glass-ceramic: Processing and Properties. *Introduction to Bioceramics*, eds. L.L. Hench and J. Wilson, World Scientific Publishing Co., London and Singapore, 89–103, 1993.

19. T. Kobuko, *Introduction to Bioceramics*, eds. L.L. Hench and J. Wilson, World Scientific Publishing Co., London and Singapore, 75–88, 1993.

20. H.R. Stanley, M.B. Hall, A.E. Clark, C.J. King III, L.L. Hench and J.J. Berte, Using 45S5 Bioglass Cones as Endosseous Ridge Maintenance Implants to Prevent Alveolar Ridge Resorption: A 5-year Evaluation. *Int. J. Oral Maxillofac. Implants*, **12**, 497, 1997.

21. H.R. Stanley, A.E. Clark and L.L. Hench, Alveolar Ridge Maintenance Implants. *Clinical Performance of Skeletal Prostheses*, eds. L.L. Hench and J. Wilson, Chapman & Hall, London, 255–269, 1996.

22. W. Cao and L.L. Hench, Bioactive Materials. *Ceramics International*, **22**, 493–507, 1996.

23. L.L. Hench and J. Wilson, Bioactive Glasses and Glass-Ceramics: A 25 Year Retrospective. *Bioceramics: Materials and Applications*, eds. G. Fischman, A. Clare and L.L. Hench, Ceramic Transactions, Vol. 48, American Ceramic Society, Westerville Ohio, 11–22, 1995.

24. T. Yamamuro, J. Wilson and L.L. Hench, *Handbook of Bioactive Ceramics: Bioactive Glasses and Glass-Ceramics*, Vol. I, CRC Press, Boca Raton, Florida, 1990.

25. T. Yamamuro, J. Wilson and L.L. Hench, *Handbook of Bioactive Ceramics: Calcium Phosphate and Hydroxyapatite Ceramics*, Vol. II, CRC Press, Boca Raton, Florida, 1990.

26. W. Bonfield, J.A. Bowman and M.D. Grynpas, Composite Material for Use in Orthopaedics, US Patent 5, 017, 627, 1991.

27. W. Bonfield, Hydroxyapatite Reinforced Polyethylene as an Analogous Material for Bone Replacement. *Bioceramics: Materials Characteristics Versus In Vivo Behavior*, eds. P. Ducheyne and J.E. Lemons, (Annals of the New York Academy of Sciences), **523**, 173–177, 1988.

28. M. Wang, W. Bonfield and L.L. Hench, Bioglass/High Density Polyethylene Composite as a New Soft Tissue Bonding Material. *Bioceramics 10*, eds. J. Wilson, L.L. Hench and D. Greenspan, Elsevier Science, Oxford, 1995.

29. J. Huang, M. Wang, I. Rehman, J. Knowles and W. Bonfield, Analysis of Surface Structures on Bioglass/Polyethylene Composites In Vitro. *Bioceramics 10*, eds. J. Wilson, L.L. Hench and D. Greenspan, Elsevier Science, Oxford, 1995.

30. R.L. Orefice, G.P. LaTorre, J.K. West and L.L. Hench, Processing and Characterization of Bioactive Polysulfone-Bioglass Composites. *Bioceramics 10*, eds. J. Wilson, L.L. Hench and D. Greenspan, Elsevier Science, Oxford, 1995.

31. R. Li, A.E. Clark and L.L. Hench, An Investigation of Bioactive Glass Powders by Sol-Gel Processing. *J. Appl. Biomat.* **2**, 231, 1991.

32. M.M. Pereira, A.E. Clark and L.L. Hench, Homogeneity of Bioactive Sol-Gel Derived Glasses in the System $CaO-P205-SiO_2$. *J. Mat. Synthesis Process.* **2**(3), 189, 1994.

33. L.L. Hench and R. Orefice, Sol-Gel Technology. *Kirk–Othmer Encyclopedia of Chemical Technology*, Fourth Edition, J. Wiley & Sons, New York, **22**, 497–528, 1997.

34. L.L. Hench and J.K. West, The Sol-Gel Process. *Chem. Rev.* **90**, 33, 1990.

35. K.D. Lobel and L.L. Hench, In Vitro Protein Interactions with a Bioactive Gel-Glass. *J. Sol-Gel Sci. Technol.* **7**, 69, 1996.

36. L.L. Hench, Life and Death: The Ultimate Phase Transformation. *Thermochimica Acta*, **280/281**, 1, 1996.

37. L.L. Hench, Bioceramics and the Origin of Life. *Biomed. Mat. Res.* **23**, 685, 1989.

INDEX

书　　名：Materials Science and Materials Engineering

作　　者：D.W.Pashley (ed.)

中 译 名：材料科学和材料工程

出 版 者：世界图书出版公司北京公司

印 刷 者：北京世图印刷厂

发　　行：世界图书出版公司北京公司 (北京朝内大街 137 号　100010)

联系电话：010-64015659, 64038347

电子信箱：kjsk@vip.sina.com

开　　本：大 32　　印　张：8

出版年代：2003 年 6 月

书　　号：7-5062-5938-9/ O・357

版权登记：图字: 01-2003- 3123

定　　价：38.00 元